WESTERN WATER A TO Z

WESTERN WATER A TO Z

The History, Nature, and Culture of a Vanishing Resource

ROBERT R. CRIFASI

UNIVERSITY PRESS OF COLORADO • *Denver*

Published by University Press of Colorado
1624 Market St Ste 226
PMB 39883
Denver, Colorado 80202-1559

Printed in the United States of America

The University Press of Colorado is a proud member of
Association of University Presses.

The University Press of Colorado is a cooperative publishing enterprise supported, in part, by Adams State University, Colorado State University, Fort Lewis College, Metropolitan State University of Denver, University of Alaska Fairbanks, University of Colorado, University of Denver, University of Northern Colorado, University of Wyoming, Utah State University, and Western Colorado University.

∞ This paper meets the requirements of the ANSI/NISO Z39.48-1992 (Permanence of Paper).

ISBN: 978-1-64642-327-9 (paperback)
ISBN: 978-1-64642-328-6 (ebook)
https://doi.org/10.5876/9781646423286

Library of Congress Cataloging-in-Publication Data

Names: Crifasi, Robert. R., author.
Title: Western water A-to-Z : the history, nature, and culture of a vanishing resource / Robert R. Crifasi.
Description: Louisville : University Press of Colorado, [2022] | Includes bibliographical references and index.
Identifiers: LCCN 2022038159 (print) | LCCN 2022038160 (ebook) | ISBN 9781646423279 (paperback) | ISBN 9781646423286 (epub)
Subjects: LCSH: Water use—West (U.S.)—Handbooks, manuals, etc. | Water use—West (U.S.)—History. | Water conservation—West (U.S.)—Handbooks, manuals, etc. | Water conservation projects—West (U.S.)—Handbooks, manuals, etc. | Water-supply—West (U.S.)—Handbooks, manuals, etc. | Water resources development—West (U.S.)—Handbooks, manuals, etc.
Classification: LCC HD1695.W4 C76 2022 (print) | LCC HD1695.W4 (ebook) | DDC 333.9100978—dc23/eng/20221125
LC record available at https://lccn.loc.gov/2022038159
LC ebook record available at https://lccn.loc.gov/2022038160

COVER ILLUSTRATIONS, CLOCKWISE FROM TOP LEFT: photo by Joshua Brown/Unsplash.com; photo by Dennis Dahl, image courtesy of the New Mexico History Museum, Santa Fe; photo by Carol M. Highsmith, courtesy of the Library of Congress; photograph by Maria J. Avila; *Rocky Mountain News*, Denver Public Library, Special Collections; public domain photo by Dale Kolke, courtesy of the California Department of Water Resources; photograph by Andrew Pernick, public domain image courtesy of the US Bureau of Reclamation; public domain image from Wikimedia Commons; Public domain image courtesy of NASA.

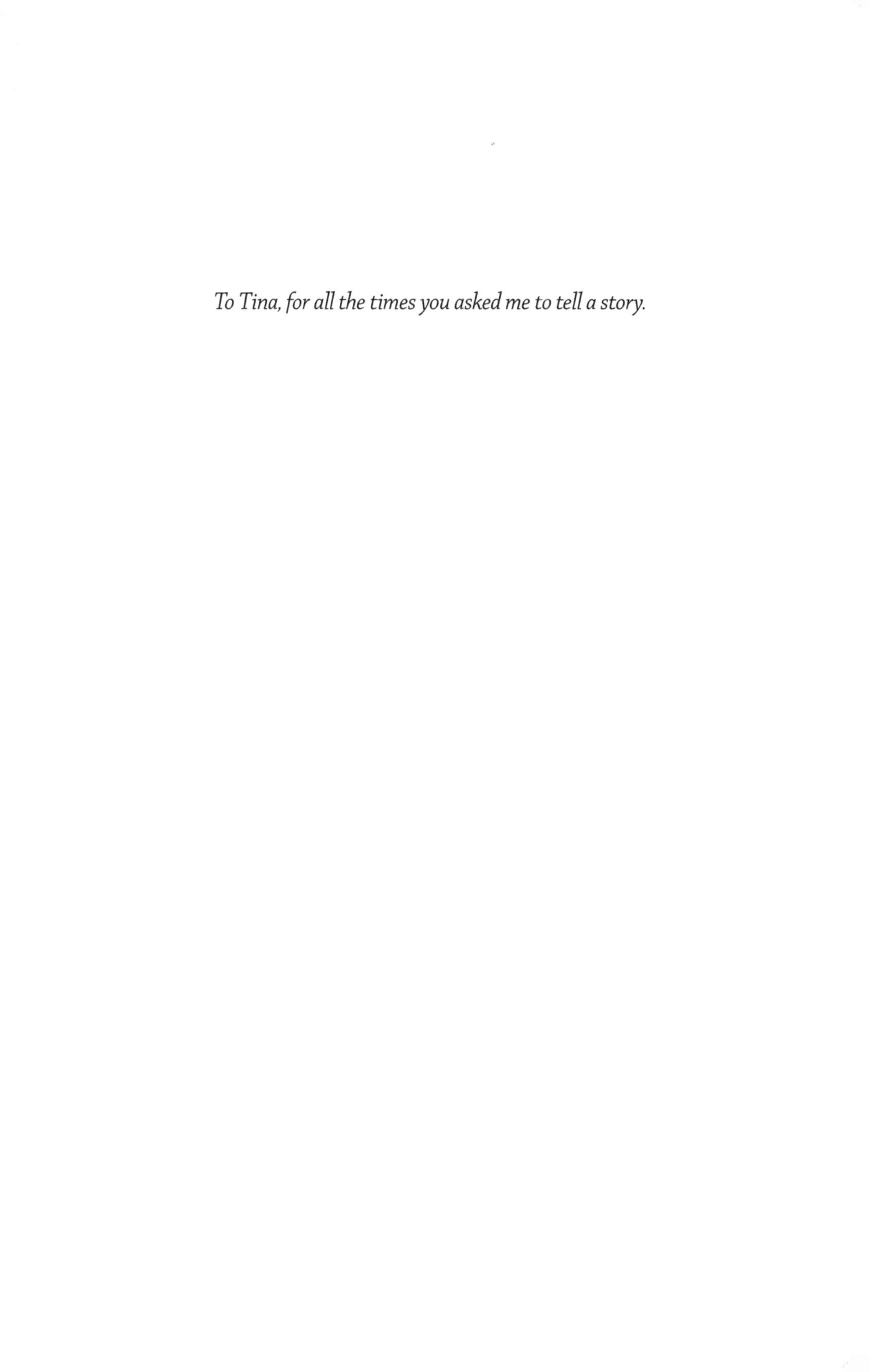

To Tina, for all the times you asked me to tell a story.

CONTENTS

PREFACE

Years ago while I was in graduate school, I had the opportunity to join the University of Colorado Adventure Program for a raft trip down the Dolores River in southwest Colorado. I was in a paddle raft, and we were fast approaching Snaggletooth, one of the more (in)famous rapids in the West. Ahead of us, the gear boat made it through the chute at the start of the rapid, but then momentarily lodged on the Snaggletooth, a gnarly pointed dagger of a rock located in the dead middle of the river. At the time, Snaggletooth, a Class IV rapid named after its namesake rock, was known for tearing holes in the rafts of even the most experienced oarsmen and women. We were next, and as we plunged into the chute, someone with a flailing paddle hit me in the forehead as I lurched forward in the raft. Bloodied but exhilarated, I saw Snaggletooth flash by us to our left.

At the time, I was only vaguely aware that I was floating on a doomed river. For years, Colorado Congressman Wayne Aspinall promoted a pet dam on the Dolores, a river within his district. With Aspinall's backing, Congress authorized the Dolores Project as part of the Colorado River Basin Act of 1968, which helped fund the earlier 1956 Colorado River Storage Act. Although the economic feasibility of the Dolores River Project was dubious from the start, Aspinall persisted with the project and finally saw it built with the construction of the McPhee Dam and its associated facilities.[1]

Congressman Aspinall advanced the Dolores Project to provide water to agriculture, and it irrigates a little over 61,000 acres across almost 1,200 farms. The water goes to alfalfa, oats, pasture land, and corn silage for livestock feed. Additionally, the towns of Towaoc on the Southern Ute Indian Reservation and Cortez in southwest Colorado receive water from the project. Still, without providing any details, the US Bureau of Reclamation reports that the "Dolores Project has provided accumulated actual benefits of $2,000 between 1950 to 1999."[2] Meanwhile, the Bureau of Reclamation spent approximately $565 million to build and another $2.8 million or so in annual costs for operation, maintenance, and repairs.[3] Although I vigorously endorse local agriculture, providing water to Indigenous communities and rural towns, I cannot understand why our society would agree to inundate a scenic desert canyon for this kind of return.

While I was rafting the Dolores River rapids, work on the McPhee Dam upstream from us was ongoing. The US Bureau of Reclamation broke ground for McPhee Dam in September 1977, and it took twenty-one years to complete the work. Unfortunately, McPhee Dam flooded a long section of the Dolores River and drastically modified its hydrology. Because McPhee Dam captures so much of the Dolores River's flow, these days there is only enough water available for raft trips like mine when exceptional snowpack in the river's headwaters yields enough water to fill McPhee and provide excess for rafting. Even then, one can expect a shortened boating season because most water goes into storage before any becomes available for river recreation. In addition to all but eliminating a rafting industry on the Dolores, the reservoir flooded Ancestral Puebloan ruins, rock art, and burial grounds, and the project to recover some of this material was the subject of the largest ever archaeological salvage program in US history. Another casualty to the Dolores River Project was the fishery downstream from the dam. Although the Bureau of Reclamation and the Colorado Division of Wildlife has agreed to maintain minimum flows for sport fish below the reservoir, the present fishery is nothing like the predam days.

In thinking about the Dolores River, it seems to be a microcosm of water issues found across the western US. Here, history, nature, and culture all become entangled. Seemingly disparate actors and ideas collide: some are winners, some are losers, and everything changes. Making difficult trade-offs appears the rule and not the exception. Each river has its own unique story, but certain things keep cropping up if you perform a grand survey across the West. All these entanglements make me think of a professor I once had who would often end an explanation about a complex subject by saying, "Clear as mud, no?" Maybe, but perhaps by writing *Western Water A to Z* I can help make this often tricky subject a bit less murky.

ACKNOWLEDGMENTS

I once read a description of book acknowledgments as a "gratuitous supplement to the dedication page." Ouch, I certainly don't see it that way. Let me make it clear: as I grovel with appreciation at the feet of those who have helped make this book see the light of day I do so with utmost sincerity. Although writing is largely a solitary endeavor, it seems that everything else about it requires the support of a small army of selfless people who directly or indirectly assist the writer. It is to all of these people that I say an unambiguous thank you.

Western Water A to Z simply would not exist without a number of amazing archival resources that store remarkable information and photographs regarding Western water. I wish to give a shout out of gratitude to the dedicated archivists and librarians at the following institutions: The National Archives and Records Administration in Broomfield, Colorado, and in Washington, DC; the New Mexico Digital Collections at the University of New Mexico Libraries; the New Mexico History Museum; the New Mexico Museum of Art; the Los Angeles Public Library; the Huhugam Heritage Center, Gila River Indian Community; the US Bureau of Reclamation; the Library of Congress; the Central Arizona Project; the Denver Public Library; the California Department of Water Resources; the Metropolitan Water District of Southern California; the Smithsonian Institution Archives; the Oregon Historical Society; the Washington State Archives; the Water Archives at Colorado State University; the University of Arizona Laboratory of Tree Ring Research; the Utah Historical Society; the University of Colorado Natural History Museum; the US Geological Survey; the Helga Teiwes Arizona State Museum; the US Fish and Wildlife Service;

the University of Wyoming American Heritage Center; the Nixon Presidential Library and Museum; the US Army Corps of Engineers; the Imperial Irrigation District; the North Dakota Atmospheric Resource Board; and the Northern Colorado Water Conservancy District.

A number of people have graciously provided photographs, images, and other materials that have greatly strengthened this book. Thank you, Matthias Koddenberg, for providing the wonderful *Over the River* rendering from the estate of Christo and Jeanne-Claude. Terrence Moore shared what is perhaps the finest photograph of Ed Abbey that has ever been made. David M. Meko kindly provided an updated version of the Stockton Chart. Mike Cecot-Schearer provided a beautiful image of a beaver foraging on the Green River. The artist Isaac Murdoch generously allowed me to reproduce his influential *Water is Sacred* poster. Mr. Murdoch, good luck with your efforts. Justice Greg Hobbs of the Colorado Supreme Court allowed the Ditch Project in Boulder, Colorado, to photograph the 1922 Colorado River Compact map that appears in this book. Joseph P. Skorupa of the US Fish and Wildlife Service and Theresa S. Presser of the US Geological Survey generously provided me with publication-quality photographs demonstrating the impacts of selenium poisoning of waterfowl.

At an academic press, the publisher typically engages several anonymous reviewers to help determine whether a submitted manuscript is worthy of publication as a book. These reviewers selflessly read a manuscript, make important observations, raise salient questions, and recommend insightful revisions without the expectation of ever being identified to the author or receiving recognition for this effort from the public. When the University Press of Colorado accepted this book for publication, I requested permission from the reviewers to waive their anonymity so that I might acknowledge their remarkable generosity. I wish to express heartfelt thanks to Elizabeth A. Koebele at the University of Nevada, Reno, Kathleen Kambic at the University of New Mexico, and Tom Cech at Metropolitan State University, Colorado, for their selfless effort. Their careful read and astute suggestions helped make this book a far better work. In particular, I value their many constructive and challenging recommendations. I took all of these suggestions to heart, and I believe that the reader is the ultimate beneficiary.

Thank you, Nina Bowen, for reviewing the Avanyu and water glyphs sections. I also wish to thank Joe Brame, Jim King, Pam King, Steven Acerson, Diana Acerson, Darlene Koerner, and Tim Sweeny for showing me the way to various rock art sites. Katja Friedrich is one of our top weather modification and cloud seeding experts. Thank you, Katja, for graciously reviewing my writing on cloud seeding and pointing me toward various references that assisted me in improving this section. Katy Barnhart, I appreciate your providing helpful feedback on the sections discussing geomorphology. It is always beneficial to have a fellow geologist challenge one's assumptions. Thank you, Charlotte Steinhardt, for your early interest in this project. To Rachael Levay, at the University Press of Colorado, I especially appreciate the enthusiasm, insightful recommendations, and overall support you provided me throughout this project. Thank you, Laura Furney and Jen Rogers, for making valuable editorial suggestions that considerably strengthened this work. I greatly appreciate the University Press of Colorado and the great staff there for their assistance.

Because the idea for this book came to me on various raft trips and road explorations, I want to thank my circle of friends who have joined me on these expeditions and for making each one of them a rewarding and memorable experience. I hope we have many more adventures. I wish to thank Tina Tan for allowing me to use a photograph of the Salt River that she made. Over the course of my academic and professional career many people, more than I could possibly name, have supported me and helped me learn about western water and its remarkable history, ecology, and cultural constructions. To all of them I say thank you!

My wife, Tina Tan, partnered with me on the vast majority of my western adventures. Her spirit and energy seem endless. She has made my life's journey so much more rewarding that I can't imagine having done it without her.

WESTERN WATER A TO Z

PART 1

WHY WESTERN WATER A TO Z?

INTRODUCTION

Water is crucial to all that exists in the American West: cities, environment, culture, politics, agriculture, development, and imagination. Water is life. It touches everything. The idea for *Western Water A to Z* emerged on various road and raft trips when I realized that few portable sources of information exist about western water even though water and water features are ubiquitous in many people's lives. On those adventures, our evening conversation often wound its way to water, and we ended up talking about history, the environment, and all the things that people do to the rivers that we love. Think of *Western Water A to Z* as an extension of those conversations. *Western Water A to Z* has its roots in those handy twentieth-century field guides (to birds, fish, trees, etc.) but is updated for a twenty-first-century audience. Perhaps by bringing this information together, I can help answer people's questions about rivers, water projects, the culture of water, the ecosystems water projects have created or destroyed, and the reliance of cities, farms, and industries on this critical resource.

History shapes popular perceptions, and as such, has the power to influence attitudes and opinions. Western water is no exception. I feel it is essential to include the most salient stories and provide some background regarding their origins and implications. Additionally, water often finds itself at the center of our cultural discourse, in art, cinema, or literature. These, too, play vital roles in shaping our understanding and experience of western water. Likewise, I include various larger-than-life personalities that are nearly synonymous with western water (John Wesley Powell, Elwood Mead, Floyd Dominy, among others). Their lives were intertwined with and often influenced the course of water development across the region.

Understanding western water requires a simultaneous review of both the physical and cultural systems that govern the hydrologic cycle and the distribution of water. Here people, along with their cultural and engineering creations, mingle with the natural environment to create the unique geography of water that we see in the West.

ABOUT THE SCOPE OF THIS BOOK

Writing *Western Water A to Z* confirmed a suspicion that I held when I began; that is, it's virtually impossible for a single author to tackle something as broad and diverse as a guide to western water. But I'm not trying to channel some nineteenth-century scribe and list every fact and detail that exists about this subject. The field of water is so vast that there is no conceivable way to cover everything that might be of interest to somebody. There are literally thousands of dams, tens of thousands of ditches, and countless natural water features in the western United States. To attempt to provide an exhaustive list would expand this book beyond any reasonable size and make any sane person glaze over in a stupor, or better, use it for campfire kindling.

Conversely, ignoring various monumental, or historic, structures would also diminish the value of *Western Water A to Z.* To strike a balance, I've included crucial projects that I believe any serious student of western water should know about. And as part of this guide, I've included descriptions and images of structures that aid in understanding the region. Likewise, there are literally dozens of terms listed in various water, geomorphology, and ecology glossaries that I do not include here. I incorporate some essential ones but omit those outside the scope and purpose of this book. If you are interested in delving deeply into the meaning of the more esoteric terms, pick up the appropriate text or take the relevant college course. Also, I include information about Indigenous water use, such as their canals and ritual beliefs (such as the Avanyu deity and the water glyphs) because most Euro-Americans almost completely neglect traditional knowledge in conversations about western water. This Indigenous knowledge is valuable and deserves a place within the discourse of western water. If you are an expert, and there are many of you, you will immediately find many places to nitpick at subjects that I only superficially address. Undoubtedly, others would incorporate specific projects relevant to them, and leave others off, or include terms I omit. Please do accept my apologies for any oversights, omissions, and abbreviations that bother you. If you find an error, please let me know! I will strive to make corrections and improve future editions. With that mea culpa, I hope you will find *Western Water A to Z* useful, entertaining, and thought-provoking.

THE ORGANIZATION OF *WESTERN WATER A TO Z*

In writing *Western Water A to Z*, I've attempted to reimagine what a guidebook can do. I always found the traditional field guides truly handy for identifying mushrooms and trees, but they never quite satisfied my curiosity about the species they discussed. My goal then, in writing *Western Water A to Z*, was to provide information about key subjects in western water but to go beyond cold factual descriptions and give you a sense of how things work and why, what makes them important to us, and how all these things are interconnected. Otherwise, one might just look up the definition of *headgate* or *alluvium* without learning their broader significance. By the way, speaking of traditional field guides, I will posit that most people don't actually *enjoy* reading the *Field Guide to Western Trees* and just use it as a resource akin to a dictionary. If only the *Field Guide to Western Trees* told us how Indigenous people used spruce trees or how this majestic tree was critical in nineteenth-century shipbuilding, perhaps one might get motivated to learn even more about spruce trees or look forward to turning the page and reading the next entry.

I also tried to describe each subject succinctly in about one page and provide one or two striking photographs. By writing *Western Water A to Z* this way, I harken back to the style of the classic field guide. There is just enough information to give you an overview of each subject, but not so much as to overwhelm you with details. In other words, just the facts. Even though each subject description stands alone as a topic, the more you read, the more you will see connections between topics. Not only do various threads connect the people whom I profiled but the landscape, culture, history, and physical environment all weave together into an intricate and beautiful tapestry that is western water. As another departure from traditional field guides, I include a few personal stories and anecdotes. This is intentional and I think necessary for reinventing the guidebook genre. Most of all, by relaying some of my experiences, you will see that we all share parts of this story, and that we're not just armchair pundits considering western water from afar.

Studying western water is a multidisciplinary endeavor. People generate knowledge about water in many different political and intellectual arenas. Each kind of knowledge is valuable and essential for gaining a better understanding of the West's water. Each source of knowledge informs the broader conversation. Anyone who fails to give equal weight to traditional knowledge, the arts, history, the physical sciences, politics, and the law does so at their peril. When I began this book, I thought about organizing it around sections about the arts, traditional knowledge, history, and so on. But the more I thought about it, by keeping subjects mixed along the A to Z format of part 2, I hoped that this strategy would nudge folks towards thinking about western water from multiple perspectives.

Somewhere in my mind, I am hoping that *Western Water A to Z* will find its way into a river guide's library of books floating down the Grand Canyon to answer participants' questions about water. Or, perhaps it will end up in the back seat of a cross-country traveler's car who might be wondering what all those water structures are that he or she sees along the way.

ABOUT THE PHOTOGRAPHY

Photographs have a unique ability to convey information and emotion about the world not easily captured in words. The philosopher Roland Barthes wrote how old photographs possess more authority than a drawing or engraving, that photographs provide "a certainty that such a thing had existed: not a question of exactitude, but of reality."[4] It is this essence that makes viewing old western photographs so rewarding. It shows us that a great canyon once existed where a reservoir now sits, or that groundwater was once so abundant that it could spray into the air. As a case in point, William Henry Jackson's photographs of geysers and hot springs helped secure congressional support for making Yellowstone America's first national park. Likewise, dam proponents often used photographs to promote reservoir projects. Conversely, Eliot Porter's heartrending images of Glen Canyon in the days before it was flooded helped bring an end to the big-dam era in the United States. And because water is so central to the western experience, we are fortunate to have exceptional works in the public archives by both famous and unknown photographers that illuminate many aspects of the subject. I have attempted to assemble in this book as many excellent photographs as I could find that illustrate the topics covered here. Also included are many images that I made to capture specific features or ideas when I could not locate a better archival photograph. Intentionally, *Western Water A to Z* is equally a work of photography as it is of water. Western water development coincides closely with the rise of photography, so images play a critical supporting role in our understanding of this story. That being so, images comprise a central position in the cultural dialogue on water and deserve recognition as something more than a documentary rendering of structures.[5]

I attempted to faithfully utilize the protocols of photojournalism in processing and presenting the images in this book. Most importantly, I have never *manipulated* any of the photos presented here. Nowhere have I moved or changed actual pixels (except removing sensor dust) in the images. However, I have *processed* the images, such as cropping, dodging and burning, converting to black and white, and have performed various conventional toning and color adjustments so that the reader sees the highest quality photographs possible.

Figure 1.1. This US Bureau of Reclamation publicity photo of the Shoshone Dam in Wyoming has an image of the US Capitol superimposed on it to illustrate the magnitude of this early reclamation project. By using photographs such as this, its boosters tried to imprint the aura of American might on the structure. Photograph courtesy of the US Bureau of Reclamation.

Figure 1.2. Eliot Porter used his photography to help sway public opinion against big dam projects in the United States. Although environmentalists were unsuccessful in protecting Glen Canyon, Porter's images ultimately contributed to the political consensus that limited further dam construction. Shown here is Eliot Porter's image *Green Reflections in Stream, Moqui Creek, Glen Canyon, Utah, September 2, 1962*, dye transfer print 10½ x 8¼ in. Collection of the New Mexico Museum of Art. Gift of Eliot Porter 1988 (1993.3.28). © Amon Carter Museum of American Art, Fort Worth, Texas. Photo by Blair Clark.

In a few instances, I came across some exceptional but faded color photographs in the archives. Here I faced a quandary: I had found a great photo, except that it was seriously worse for the wear with time. When this occurred, I used the digital tools available to me to restore some of the original colors and reverse the fading so that the image appeared (at least in my mind) to align with what the original photographer may have expected when he/she received the photos back from the printer. Of course, I truly cannot hope to know what they expected from the print. Still, the archival images that received this treatment were vernacular and produced as illustrations in reports and not presented as fine art (even though many are great photos). Therefore, I do not think I subverted or reinterpreted the intent of the photographer. But by processing them, I believe that these particular images can now be viewed anew and by a wider contemporary audience.

Let me share one last observation I made after spending extensive time in the archives. It seems that the older black-and-white images have tended to withstand the test of time far better than more recent color photos. I hoped to include more color images from the 1960s and 1970s, but so many had faded to the point of uselessness. So, unless the image was significant, I tended to use older photos because they simply looked better. Archival color photographs from the 1960s and 1970s will be lost to us as the dyes within the images continue to degrade unseen in archival collections.

WHAT IS THE AMERICAN WEST?

The historian Wilbur Jacobs once described the difficulties of defining "the West." "As you move back and forth," Jacobs explains, "the West as a place floats on the map, almost like a puddle of mercury. The sub-puddles, spinning around, have so many socioeconomic, political, environmental, and cultural eddies that they are almost impossible to control when we try to write a coherent account."[6] With Jacob's wisdom in mind, it's tempting to stop right there, move on, and dodge the question of "what is the American West?" But I will resist this urge because confronting it will help add perspective to the topics and questions concerning western water that follow.

Central to any conversation about the West is how do we describe its remarkably diverse people? What different groups call themselves—and what they want others to call them—is especially critical for having respectful and informative conversations. I recognize that there are many different terms for describing various groups, and that terminology appears in almost constant flux. Many people consider words formerly used for some groups as out-of-date or even insulting if used today. In *Western Water A to Z*, I have endeavored to use consistent, deferential terms when discussing various populations. For these reasons, unless the words *Indian* or *Native American* appears in a Tribal name, historical event, or quote, I use *Indigenous* to refer to the original inhabitants of the West. Since this book discusses the water systems of northern New Mexico, and because these systems have roots extending back to Spain, I use the term *Hispanic* to inclusively describe all of the people who can trace their origins to Mexico, Central America, and Spain. When I use the word *Euro-American*, I apply this in the broadest sense possible. I include all American citizens (regardless of race or continent of origin) and the European colonists who came to North America and migrated west. Additionally, I feel the terms *settler* and *pioneer* are quite dated as they do not accurately reflect the reality that the Euro-Americans colonized lands occupied by others. For this reason, I use *colonist* and not *pioneer* or *settler*.

Most archaeological evidence points to the peopling of North America at about fourteen thousand years ago. However, that date remains controversial, and more recent findings suggest significantly earlier occupations.[7] Without debating archaeology, one can confidently state that much of North America contained substantial populations of Indigenous people for a minimum of fourteen thousand years. For these people, there was no American West. To each tribe, wherever we might place them on a modern map, they lived at the center of their world. To the descendants of these Indigenous people, the idea of the American West was an imposed foreign concept. As Tohono O'odham Nation Chairman Ned Norris Jr. put it, "The O'odham have lived in Arizona and northern Mexico since time immemorial. We experienced the impact of a border through our lands in 1854—a border that was drawn without regard for our history, our original territory boundaries, or our sovereign rights."[8]

When Hernán Cortés and the conquistadores conquered the Aztecs in 1519, they established a Spanish colony in the heart of Mesoamerica, with Mexico City as their capital and colonial center. Like the Indigenous people before them, the Spaniards possessed no concept of an American West. To them, the western half of North America was part of the Spanish possessions they claimed as New Spain. The Spaniards divided up what we now know as the western United States into the provinces Alta California, Nuevo Mexico, Alta Louisiana, Baja Louisiana, and Texas. For the Spanish, and the Mexicans after 1821, the idea of a western United States was seditious.

What eventually came into focus as America's northern and southern boundaries resulted from a series of overlapping conquests of North America.[9]

Beginning with Cortés, Spanish (and after 1821 Mexican) incursions and conquests gradually moved the frontier between Spanish and various Indigenous tribes north from Mexico. Some two hundred years later, British, French, and eventually American incursions and conquests gradually pushed the western frontier between the Europeans and Indigenous inhabitants westward, first across the Appalachians, then the Mississippi, and finally to the Pacific. American expansion inevitably led to conflict with Mexico, giving rise to the contemporary southern border with Mexico with the signing of the Treaty of Guadalupe Hidalgo in 1848 and the Gadsden Purchase of 1854. Territorial conflicts and competition in North America between the United States and Great Britain, beginning with the American Revolution, were finally resolved by signing the Oregon Treaty in 1846. This treaty established America's northwestern border along the forty-ninth parallel to the Strait of Georgia, just short of the Pacific Ocean. These various treaties demarcated the north and south sideboards on what we now know as the western United States.

If we take the Pacific Ocean as the continental United State's natural western boundary, then there is one remaining and very thorny question to consider: Where on the east side of this great box does the West begin? For some geographers, the Mississippi River marks a convenient boundary. To others, the Rocky Mountain Front, that boundary between the eastern slope of the Rocky Mountains and the Great Plains, is a logical demarcation. Some might prefer a line separating the High Plains' grasslands from the forests of the Rocky Mountains. Perhaps we should think of the edge in economic terms, say along the lines of William Cronon's analysis of regional dominance that distinguishes the easterly region dominated by Chicago from the westerly area dominated by Los Angeles and San Francisco.[10] Then there is a political option: say that the West begins at the eastern boundaries of Montana, Wyoming, Colorado, and New Mexico. Finally, if you are a hydrologist—and this is a book about water—call the West's start at the one hundredth meridian, that general line of longitude that roughly marks the transition from the humid East to the arid West. That, after all, is the logic that Major John Wesley Powell used to distinguish the West from the East.[11]

It seems that defining the West is a matter of one's perspective.[12] As we can see, there is no right or wrong answer, just the one that best suits your needs. For this book, I'll define the West as a great box bounded by the Pacific Ocean on the west, the Canadian and Mexican borders on the north and south, and I will throw my lot in with Major Powell and say that its eastern margin begins at approximately the one hundredth meridian.[13]

HISTORY OF THE WEST AND ITS ENVIRONMENT: A LAND OF SCARCITY?

We often speak of the American West in terms of its aridity. And by most definitions, it is, in fact, an arid or semiarid place. Deserts and steppes dominate the western environment even though islands of wetter mountain terrain tend to grab our visual attention. It is in these deserts and steppes that we dig ditches to bring water to our agricultural land. In the East, they dig ditches to move water off of fields so that it does not drown crops. To be sure, it is wet in many of the western mountains, but relatively few people live high up. Most folks live in the valleys and on the plains where constructed waterworks make the place habitable. So, we get to the central conundrum of the American West: in this arid land, it is water that is the region's most defining trait. And if we wish to garner a deeper understanding of the West, we must get to know the natural and human-made features that are only here because of water and our use of it.

But there is also a second, subtler, conundrum at work here. Put simply, although this is a desert land, water itself is not necessarily *scarce*. Bear with me on this one. Just look at Lake Powell. Its water level might be quite a bit lower than its builders would wish it to be, but still nevertheless contains an enormous lake. The Colorado River that fills Lake Powell is fully appropriated, overused, and overloved, but even so, it remains for most of its course a grand river. Even its delta—which is often dry due to the hands of humans—is still a vast area that formed over millennia through the force of water. The problem is that our culture is one that seems blind to the reality of natural limits. What we have is more a problem of allocation than a water deficit. We treat the resource as if it is an infinite commodity. We keep engineering our waterways to squeeze more water out of them rather than do the obvious, albeit politically tricky thing, and redesign our cities and legal codes to better share what we already use and live within this land's essential environmental constraints. In writing this book, I hope that you reach the same conclusion I did a long time ago, namely, that there is a remarkable natural abundance of water in the western deserts and steppes. I believe this paradox shows that water scarcity and drought are at its core culturally manufactured phenomena. Although that may sound a bit odd at first, it means that the solutions to this scarcity lie within creating a willingness to adapt our culture and politics more than further manipulating our environment. I believe that, in the long term, we will find more success in figuring out how to share water and use it better than to mine the environment further or import it from afar.

CONQUEST

In short order, I will discuss prior appropriation water rights. However, before I can meaningfully do that, I must first address head-on the unsavory back story that made this legal doctrine (and for that matter, modern water development in the West) possible. In the nineteenth century, the United States expanded its borders west of the Mississippi River through war, treaties, and purchases with France, Mexico, and Great Britain. These actions gave the United States titles to the western lands that it could then pass to colonists. The Indigenous people had no say or standing in these proceedings even though these events dramatically impacted (and in many cases ended) their lives. Mapping by explorers, such as Lewis and Clark, John Charles Fremont, and John Wesley Powell, among others, aided in America's westward advancement. Manifest Destiny as experienced by the Indigenous people meant war, genocide, ethnic cleansing, forced treaties, and colonization. For Hispanic peoples, it meant forced assimilation into a new country. The violence, hypocrisy, and arrogance of this era is monumental.[14]

In the western territories, many Euro-American colonists actively worked to kill or remove Indigenous people who would stand in the way of their settlements. The tragic events surrounding the Sand Creek Massacre, and Colorado's colonists' underlying motivations to participate in the massacre, offer a case in point.[15] Many colonists acquired titles from the United States to the lands they were squatting after participating in this crime. For the territorial elites, the spoils were even greater as they received vast tracts of land along with the forest, grassland, water, and mineral resources as title passed from the Indigenous people to the federal government and then either directly on to the elites or indirectly to them as railroad land grants. Colorado's colonists were by no means alone acquiring their land and water this way, as the Indian wars that afflicted every western territory demonstrate. Because of this, when I discuss water management's legal framework, keep in mind that the subjugation of Indigenous people preceded Euro-American claims to water under the Prior Appropriation Doctrine.

In the aftermath of the Mexican-American War, the United States was concerned about appeasing the Mexican population that found themselves on the new border's American side. The Americans knew that their military presence was weak and had little ability to suppress potential unrest within the (former) Mexican settlements. Take the Taos Revolt of 1847. This brief insurrection led to the death of Territorial Governor Charles Bent and several other Americans. Territorial authorities swiftly dispatched US troops to Taos. A short and bloody battle ensued in which US troops killed approximately 150 insurgents. The

Americans captured the ringleaders and after a quick trial hung six of them in the Taos village square.[16] To maintain control, the United States had to assuage the local population's fears that their freedom and property were safe. Consequently, in 1848 as a significant concession in the negotiations formally ending the war, the Treaty of Guadalupe Hidalgo provided that "property of every kind, now belonging to Mexicans not established there, shall be inviolably respected. The present owners, the heirs of these, and all Mexicans who may hereafter acquire said property by contract, shall enjoy with respect to it guarantees equally ample as if the same belonged to citizens of the United States."[17] This provision formed the basis for recognizing Hispanic land grants and water rights in the American Southwest. Territorial legislatures later recognized and adopted many aspects of water law from the Mexican era such as rules governing the operation of the traditional ditches that they call *acequias*.

The history of American conquest and colonization leads to an odd contradiction in writing about western settlement. On the one hand, many Euro-Americans participated in brutal wars of conquest. On the other hand, many of these same colonists built vibrant cities, farms, and universities. Moreover, Euro-American migrants copied the Hispanic and Indigenous agricultural and irrigation practices, adopted some of the language to describe the landscape, and then relied on Hispanic and Indigenous labor to work their new farms and ranches. Thus, we see the same people simultaneously criticized for their colonial abuses and heralded for their nation building.[18]

PRIOR APPROPRIATION AND THE INSTITUTIONS OF WATER MANAGEMENT

By putting Western development in context of nineteenth century events, I now turn to the American colonist's activities in managing and developing water resources. When early Euro-Americans came to the West, they were a practical bunch. On arriving in places like Colorado or Utah, they often built their ditches first, and then attended to other matters like constructing houses, filing for water rights, or organizing ditch companies. The reason they did this was simple: without ditch water, they couldn't grow crops, and without crops, they would have nothing to eat come winter. They could figure out the legal niceties after they secured a stable food supply.

Once the colonists had a little spare time, they started to work out how to manage the land and water they had stolen from the Indigenous people. The first Euro-Americans were squatters on the public domain and Indigenous territory.

The Homestead Act did not exist when the earliest Euro-Americans arrived. And many of the treaties that ceded Indigenous land to the Americans were not yet inked. These first colonists imported the legal customs for managing water that they learned in the eastern United States.

Initially, the legal framework for managing water that the colonists used was known as the Riparian Doctrine. This doctrine allowed landowners bordering a water body to acquire certain rights to use the water. As legal scholar David Getches once put it, "Each landowner bordering a waterbody may make reasonable use of the water on the same land if the use doesn't interfere with reasonable uses of other riparians." Twenty-nine states still use the Riparian Doctrine in one form or another. Most scholars believe that the Riparian Doctrine came to the United States from English common law, but there are good arguments for French origins too. Regardless of the specific European country where the Riparian Doctrine originated, this doctrine became the primary way for administering water in the United States in the aftermath of the American Revolution.[19]

At first, some western territories adopted the Riparian Doctrine. But in the massive influx of Euro-Americans to California during the 1849 Gold Rush and then to Colorado in *its* Gold Rush, the migrants found the Riparian Doctrine ill-suited in the arid West. A system that limited water rights to only those people that bordered streams could drastically constrain development. Under this system, one person could acquire stream-front property and cut everybody else off from water. The old adage that if you control the water then you control the land was certainly true in the nineteenth-century West. Colonists feared that water monopolies would strangle economic growth before it could even get started. So, when the Euro-Americans turned their attention to writing the laws for gold mining and water distribution, they rejected the Riparian Doctrine as unsuitable for their needs. What emerged was the Prior Appropriation Doctrine.

Essentially, the Prior Appropriation Doctrine is a set of ideas that westerners formalized into law over the last one hundred and fifty years. Perhaps the most critical concept in the doctrine is the notion of first in time, first in right. This notion comes from the California 49ers, who said that the first person to work a mining claim had the most senior right to the site. It's kind of like getting in line early to get a ticket for a front row seat at your favorite concert. When farmers adapted this idea to water use, it came to mean that the first person to claim water had the highest priority to the stream. The second person had the second most senior right, and so on. From a practical standpoint, it means that as a stream dries up, the final person to lay claim to water is the first one that the local water commissioner shuts off. And the one that owns the earliest water right is the one that is the last to be shut down due to dry conditions.

Another idea embodied in the Prior Appropriation Doctrine is the "use it or lose it" principle. Use it or lose it means that you must keep up your water use, and if you fail to use water, a court may find that you have abandoned it, thereby moving junior rights up in priority. Yet another part of the doctrine is the notion of beneficial use, which compels users to manage their water carefully so as not to produce excessive wastewater so that other users can get some too. And importantly, the Prior Appropriation Doctrine also allows water to get diverted or transferred from one watershed to another.

As Euro-American colonists worked out the ideas for what became the Prior Appropriation Doctrine, it became clear that they required documentation to establish the relative priority of rights on each stream. It was in Colorado that the courts and legislature began fleshing the components of water rights and the paperwork for recording them. To this end, Colorado authorized the first "stream adjudications," the court proceedings that recorded details about water rights, including the relative priorities of one right to another.

These water rights are the legal documents that permit a user to divert water from a natural source. In the West, water rights specify the location where a user may divert water, the source of the water, how much water they can divert, the location of use, and the appropriation date. Under the Prior Appropriation Doctrine used throughout the West, the earlier the appropriation date, the more valuable the right tends to be. Many states including Colorado reserve the ownership of water to the people of the state, but then grant a right to use that water. Unlike other real estate, someone can forfeit their water rights if they let the rights fall into disuse.

Notably the value of water rights is rarely taxed at the time of a sale, even though water rights often make up a sizable portion of the cost of a property. Moreover, water rights owners can sell the rights separately from the land. Even so, the new owner must then secure permission to change the type of use allowed, timing of use, or location where the water is applied before actually being able to use the water at a new site.[20]

Significantly, the Prior Appropriation Doctrine has its origins in agrarian anticorporate populism. In the nineteenth century, many colonists were very concerned that large corporations or wealthy individuals might monopolize water resources and thereby deny whole regions of the water supplies needed for development. States such as Colorado declared that water is the property of

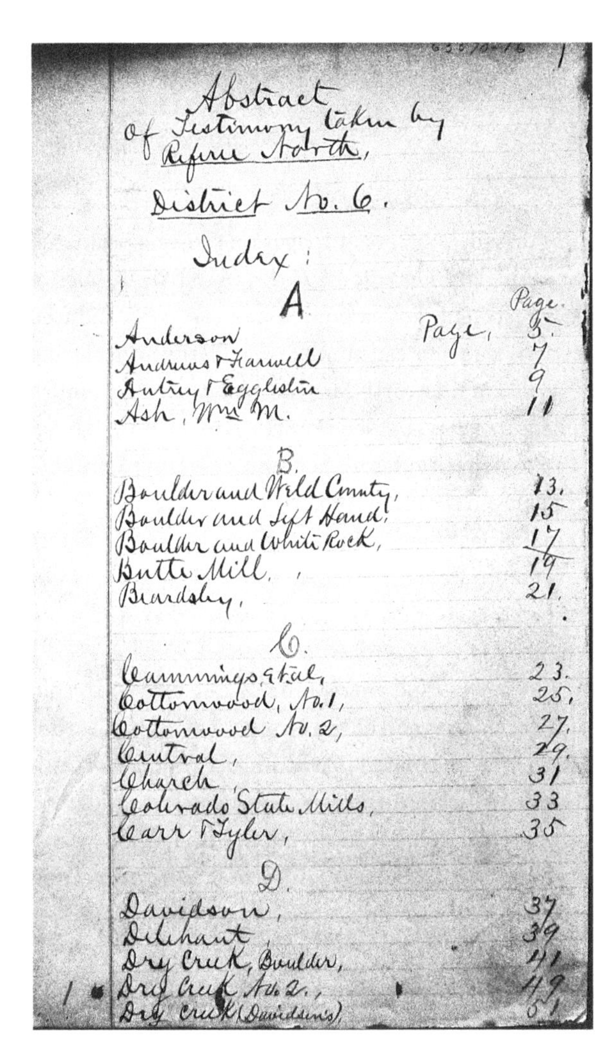
Abstract
of Testimony taken by
Referee [illegible],
District No. 6.

Index:

A

	Page.
Anderson	Page, 3.
Andrews & Farwell	7
Autrey & Eggleston	9
Ash, Wm M.	11

B.

Boulder and Weld County,	13.
Boulder and Left Hand,	15
Boulder and White Rock,	17
Butte Mill,	19
Beardsley,	21.

C.

Cummings et al,	23.
Cottonwood, No. 1,	25,
Cottonwood No. 2,	27.
Central,	29.
Church,	31
Colorado State Mills,	33
Carr & Tyler,	35

D.

Davidson,	37
Delehant,	39
Dry Creek, Boulder,	41
Dry Creek No. 2.,	49
Dry Creek (Davidsons)	51

Figure 1.3. Recording and documentation of water rights throughout the West is critical in maintaining orderly water use. During the first years that colonists worked out all the bugs in the water rights administration system, courts took testimony and assigned the relative priorities of water rights for future administration. These court processes were known (then and now) as adjudication proceedings. The first formal adjudications in the American West took place on creeks along Colorado's Front Range. Here is the abstract for the adjudication that took place on Boulder Creek near Boulder, Colorado, in 1881. Each page of testimony recorded all manner of information, including the ditch construction date, who dug it, when they began, how much water they diverted, and so forth. Photo courtesy of the author.

the people and put a decidedly democratic socialist stamp on the constitutional and legal framework for water use that remains in place today. In other words, the legal principles for water use across the West, and in particular the economies of rural agricultural areas, depend on an inherently socialistic system for their vitality.

One twist is that a few western territories initially adopted the Riparian Doctrine but later converted to the Prior Appropriation Doctrine. Faced with reconciling two contrary legal principles, the legislatures and courts in these territories wound up blending aspects of the riparian and appropriation systems to end up with a hybrid, or "California doctrine." California, Oregon, and Washington adhere to this hybrid system.[21]

WATER DEVELOPMENT

As American colonization progressed, boosters of irrigation projects promoted their ideas with the zeal of carnival barkers. Some resorted to pseudoscience or outright lies to encourage investment. For example, Samuel Aughey induced thousands of migrants to settle arid farmsteads in the High Plains by declaring that "rain follows the plow." William Ellsworth Smythe espoused the promise of irrigation with an evangelist's passion that led untold colonists onto barren western lands in the belief that prosperity was all but assured. Dam promoters of the early twentieth century pledged that their projects were the critical ingredient that guaranteed that great cities would grow out of the desert. Engineers shamelessly compared their concrete structures to the pyramids of the ancient Pharaohs. Deserts shall become an oasis, rain would follow the plow, and independence shall spring from dependency.[22]

During the same time that promoters and entrepreneurs dug ditches or sold land, inventors, engineers, and scientists solved many practical problems that impeded water development. Inventors created machines that could economically dig ditches or efficiently irrigate fields. Engineers and scientists worked out how to measure water flows by designing gages and giving us standardized water measurements such as "acre-foot" and "cubic feet per second." They advanced many aspects of water science on subjects as diverse as evaporation and river dynamics. Others devised ways to construct even bigger dams and purify water, transport water more efficiently, or build hydroelectric power plants. It is in these numberless inventions, innovations, and improvements that the modern West emerged.

During the nineteenth and twentieth centuries, the West became connected to the eastern United States and eventually the broader world by building railroads, developing ports, and constructing superhighways. In the West, agriculture gradually shifted away from individual farmsteads that supported families to larger operations producing commodities. By the mid-twentieth century, considerable numbers of small farmers either sold out to larger operations and moved to the city or became contractors for corporatized farms. Instead of raising a few cows for a family, ranchers began raising herds to ship to metropolitan areas. The patterns of water use shifted along with these changes. Some farmers sold their water to cities or industries, and large farms acquired the water rights from small farms. Many regions began specializing in crops best suited for that area. Western cities could grow by absorbing nearby farms, ranches, and their water rights. And western cities could sustain higher population densities than local agriculture could otherwise support by importing farm commodities from other states and nations. By relying on intranational and international food supply chains, the arid West has seemingly become independent of some of the climatic constraints imposed by their local dry climate.

As the twentieth century progressed, the federal government became one of the major benefactors for western water development. The federal government also set aside vast tracts of public domain as National Parks and Monuments, Wilderness Areas, Wildlife Refuges, and Wild and Scenic Rivers. Figures such as Elwood Mead and Floyd Dominy used their influence as the commissioners for the US Bureau of Reclamation to get reservoirs such as Hoover, Grand Coulee, and Glen Canyon Dams constructed. Other agencies, such as the US Army Corps of Engineers, completed massive hydroelectric projects. Building western water projects did not occur without resistance. As directors of the Sierra Club, John Muir and David Brower fought to protect American rivers. And writers such as Ed Abbey and Norman Maclean showed Americans the value of free-flowing rivers. Overall, many thousands of Americans built ditches and dams, worked to secure clean drinking water, or strove to create parks and natural areas. We can read much of the West's history in its water.

In thinking about the course of water development, it's helpful to keep in mind all of the physical, hydrological, and biological conditions that influenced or constrained people's decisions for building water projects.[23] Geographers call the environmental factors that influence, often unconsciously, human decision-making "environmental determinism." In the broadest sense, the West's aridity pushed people toward building ditches and dams. Likewise, the physical environment dictated where people constructed farms, transportation routes, water infrastructure, and cities. And once people started building up infrastructure, existing projects often dictated where new projects would go next. For example, this "path dependency" means that once a dam gets built on a river, the planners designing the next water project must consider existing dams and all the other environmental and human factors they face.

In America, water development became a cultural metaphor for progress during the twentieth century. It also became a symbol for despoliation of the environment, corruption, and pork barrel politics. Artists, photographers, writers, and filmmakers all tilled fertile grounds in their explorations of what water means for our western culture. Much of our collective understanding of the West comes from how these artists represented the region. These representations in turn mold our perceptions of western water and our political choices governing water development.

Across the West, many water projects were conceived but never built, or built and later abandoned. These projects failed due to any number of issues, including staggering costs, the collapse of political will, insufficient water, unsupportable maintenance, siltation, unrealistic aspirations, and hubris. Projects that never left the drawing board include proposals to transport icebergs from the arctic and massive canals from the Columbia and Missouri Rivers to the cities of California and Colorado. The Bureau of Reclamation made plans to construct dams in Grand Canyon National Park and Dinosaur National Monument that never materialized due to stiff grassroots opposition, cold feet by politicians, and shifting priorities to preserve natural environments in their ancient state. Proposals for dams to slake the thirst of Denver (Denver Water's Two Forks), Las Vegas (Southern Nevada Water Authority pipeline), or High Plains farms (Narrows Dam) got scuttled for want of clear need, their exorbitant cost, local opposition, and potential environmental harm. Scattered among the wreckage of large projects are multitudes of small ditches, stock ponds, and minor reservoirs that have fallen into disuse as ranchers and farmers age, aquifers decline, or the promised benefits of the structures never materialized.

If we want to gain a more complete understanding of western water, it's important to include as many perspectives as possible. When we talk about water in the West, it's easy to shift between topics that touch on cities, environment, agriculture, art and culture, politics, development, people, colonization, history, and imagination. Having a baseline understanding of these topics increases our collective fluency for discussing western water.

PART 2

WESTERN WATER A TO Z

Figure 2.1. Here is Ed Abbey in 1986 near his home in Tucson, Arizona. Image courtesy of © Terrence Moore.

ABBEY, EDWARD

When driving around the American West, keep an eye out for the bumper stickers on beat-up old Subarus and Toyotas, and you'll invariably come across one that exclaims, "Hayduke Lives!" Insiders know this is a reference to Ed Abbey's celebrated character from the irascible book that made him famous, *The Monkey Wrench Gang*. Described by the *New York Times* as an "underground cult hero throughout the West," Ed Abbey (1927–1989) galvanized the environmental movement with his two most popular books, *Desert Solitaire: A Season in the Wilderness* and *The Monkey Wrench Gang*. In particular, *The Monkey Wrench Gang* tells the tale of a band of ecological saboteurs who roamed the West, destroying railroads, bridges, and other infrastructure that they deemed affronts to the environment. In it, Hayduke was the long-haired, Vietnam War vet turned environmental activist, who fought to save America—that is an authentic America represented by western public lands and wildlife—from despoliation at the hands of greedy capitalists. Hayduke and his cohorts' overarching dream was to run a houseboat loaded with high explosives into Abbey's ultimate symbol of western degradation, the Glen Canyon Dam. It would drain Lake Powell, the water body that Abbey derided as a "storage pond, silt trap, evaporation tank and garbage dispose all, a 180 mile-long incipient seweage [*sic*] lagoon."[24] Abbey's writings motivated many to join mainstream environmental organizations such as the Sierra Club and influenced others to abandon the mainstream and throw their lot in with radical environmental groups such as Earth First! and Greenpeace. A few, inspired by Abbey's fiction, went on to undertake actual acts of environmental terrorism. In books such as *Down the River*, Abbey firmly linked his name to river conservation, restoration, and opposition to new water projects in the West. Abbey's cries in favor of wilderness, the "idea of wilderness needs no defense. It only needs more defenders,"[25] and rants against capitalism, "One thing more dangerous than getting between a grizzly sow and her cub is getting between a businessman and a dollar bill,"[26] situated him as one of the preeminent voices during the late twentieth century to favor the environment over economic development. This, and his advocacy for free-flowing rivers—"here we discover the definition of bliss, salvation, Heaven"[27]—placed Abbey at the apex of twentieth-century western environmental literature.[28]

ACEQUIA

In the 700s the Arabs and Moors brought irrigation technology to Spain when they invaded the Iberian Peninsula. Their irrigation technology took root in Spain, and the Spanish Conquistadores, in turn, brought their arid lands technology to the Americas. Among their most important innovations for irrigating desert lands was the construction of ditches or "acequias." Derived from the Arabic *al saqiya*, *acequia* is a Spanish word meaning "water conduit" or "water carrier" that has entered English. In some places, a similar Spanish word, *zanja*, meaning "ditch," is used instead of acequia, as for example with Los Angeles's *Zanja Madre* or mother ditch. By necessity, acequia construction began with the arrival of the first Spanish colonists in 1598, and acequia management was often the primary form of government implemented in the Hispanic Southwest. Today, acequia governance frequently remains synonymous with local government in rural Hispanic communities. Then, as now, communities organized themselves to build and maintain ditches and to create rules or customs, known as *reparto*,

Figure 2.2. Here volunteers are helping with the 1984 annual cleaning of the *Acequia Madre* (Mother Ditch) in Santa Fe, New Mexico. Image courtesy of the Palace of the Governors Photo Archives, New Mexico History Museum, Santa Fe, New Mexico, Negative number HP.2014.14.71.

to allocate water. During times of plenty, water users might take as much as they wished. But during times of scarcity, the *mayordomo*, or "ditch boss," divides water according to need by rotating the times during which a user may take the water. When drought strikes, a mayordomo might further restrict water deliveries to only gardens and livestock.[29]

In the arid Southwest, the area in the American imagination that defines the very notion of water scarcity, it is water itself more than any other resource that delineates the region's character. It is no surprise then that the master ditch diverting water for the Hispanic communities is known as the *Acequia Madre*, or "mother ditch." As the name suggests, this ditch is a source of fertility and plenty for the community as a whole. To manage these ditches, *parciantes*, or water users, must supply labor in proportion to the land they own to help clean and ready the acequia for the spring water run. Today, acequia organizations find themselves under pressure as urbanization consumes formerly irrigated land, and users sell rights to cities or developers. Nevertheless, these organizations remain some of the longest continuously operating institutions within the United States. As such, they represent a standard for sustainable water use and cultural preservation.

ACRE-FOOT

In the mid-1880s John Wesley Powell and his federal Irrigation Survey coined the term *acre-foot* to describe the amount of water that would cover an acre of land one foot deep. The ungainly term gained popularity as a convenient number for describing volume. In particular, it allows one to quickly calculate how much water a farmer might need to irrigate his or her land.[30] It also conveniently shortened reports (by lobbing off many zeroes) that had previously written reservoir volumes in cubic feet. An acre-foot, the writer Ed Quillen explained, is "the amount of water an average family of four uses in one year, but this definition is too fluid; only in desert regions is it appropriate. For example, in a wet state like Minnesota, the average family of four consumes only 0.44 acre-feet of treated water in a year, and in Oregon, it's all of 0.34 acre-feet. But in dry Colorado, it's 0.93 acre-feet; arid Wyoming, 0.96; thirsty Arizona, 0.99; desert Nevada, 1.12; and parched Utah, 2.46. These figures demonstrate that treated water is unlike other commodities: the less gasoline there is available, the less people consume, but the less water there is, the more people consume." Of course, the added consumption in arid states that Quillen refers to is mostly from outdoor irrigation that is not needed in wet states such as Minnesota. But Quillen tangentially

Figure 2.3. It is easy to visualize an acre of land covered by water a foot deep. It is something altogether different to contemplate the magnitude of water stored in reservoirs across the American West. To give a sense of this scale, here is a satellite image of the 18,700,000 acre-foot (Or 814,572,008,294 cubic feet if were not for Powell's Irrigation Service) Fort Peck Reservoir on the Missouri River. When you see the lake from a satellite, the lake extends 134 miles across Central Montana and covers about 245,000 acres. Completed in 1940 by the US Army Corps of Engineers, it is the fifth-largest human-made lake in the United States. In the American West, reservoirs big and small, such as Fort Peck, store millions of acre-feet, serving agriculture, industry, and municipal water supply needs. Image source: screen capture Google Earth.

suggests an important point, namely that outdoor water use in Western urban areas is mostly an amenity—it's not essential—and therefore makes a good target for water conservation.[31]

AGGREGATE MINING

"The world is running out of sand," proclaimed the *New Yorker* magazine in an article that reported "China's swift development had consumed more sand in the previous four years than the United States used in the past century."[32] This is because sand and gravel are one of the essential raw materials for development. Home construction requires an average of more than 100 tons of aggregate, and a typical mile of interstate highway uses about 38,000 tons for every mile of lane built. As such, sand and gravel are critical commodities for development worldwide. Gravel or aggregate mining has supported urban development for decades. In the American West, aggregate mining typically takes place along floodplains in unconsolidated stream alluvium. During gravel mining, miners sometimes dig

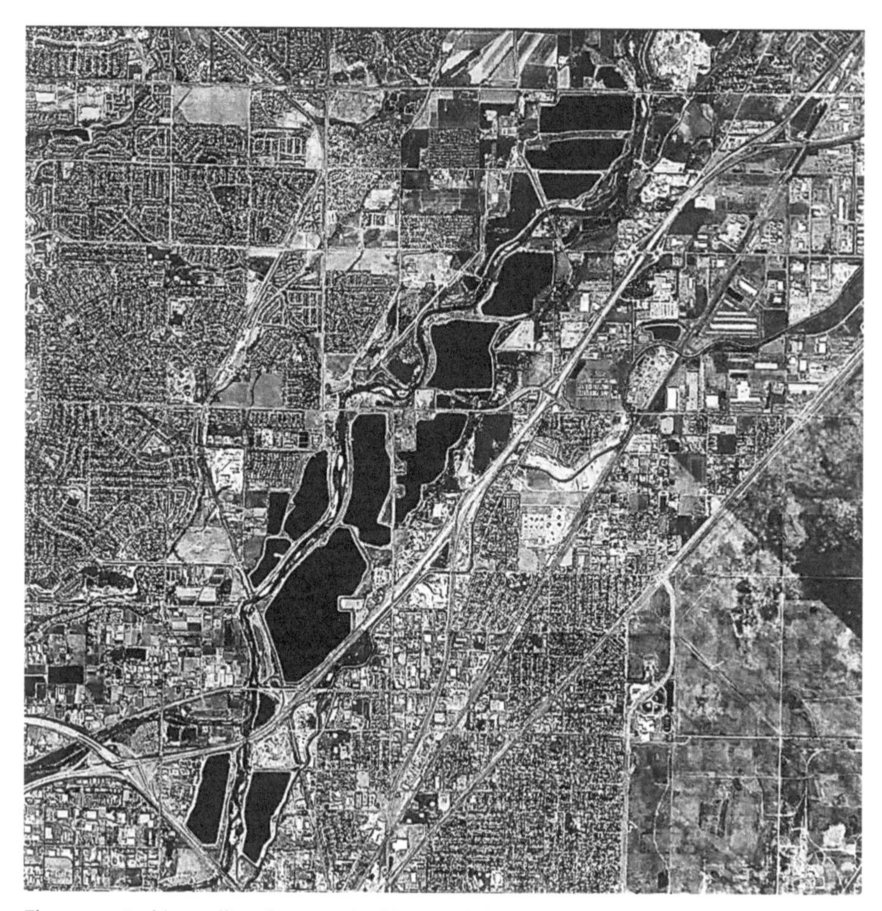

Figure 2.4. In this satellite photo north of Denver, Colorado, the dark areas are numerous gravel pit lakes and reservoirs that border the South Platte River. On Colorado's High Plains, where this photo was taken, virtually no lakes existed before active settlement. By excavating gravel pits to depths below the groundwater table, aggregate miners created conditions that led to the pits filling with water once the gravel resources were exhausted and the dewatering pumps shut off. At these sites, unique hybrid ecosystems have formed where nothing similar previously existed. Image source: screen capture Google Earth.

down below the groundwater table, which forces the mining companies to pump water from the pits to maintain production. When the aggregate is gone and the pit is abandoned, the company ceases pumping, and the former pit may fill with water to create a lake. In some areas, such as along the Front Range of Colorado, upwards of 25 percent (by surface area) of all lakes present are former gravel pits.[33] These abandoned pits often provide valuable wetland ecosystems that support fish and waterfowl habitat. However, because so much water evaporates from these lakes, it can harm senior water rights, and many states require gravel pit operators to file for water rights to replace the water lost to evaporation.

Eventually, water developers recognized the holes in the earth left by miners as a potential resource. Starting in the mid-1980s municipalities began lining former gravel pits with impermeable membranes so that they might store water and operate the sites as reservoirs. The concept became favored as a cost-effective way to create new reservoirs. Using former gravel pits for storage has expanded as an environmentally preferable alternative to damming rivers.

AGRICULTURE

We all need to eat, and agriculture is there to provide our food. Beyond the necessary consensus of need, there is little common understanding or agreement on how to best supply the essential food and fiber that our population demands, particularly in arid regions of the American West. Upwards of 90 percent of all water consumed in the American West goes to agricultural uses. That said, the diversity of water use within the agricultural sector is quite stunning. Agriculture inherently includes the cultural practices of farming, tillage, animal husbandry, cultivation, horticulture, and the various forms of working the land to grow crops and raise animals for food and fiber. Water is, of course, used to irrigate the land, but is also used in every other aspect of food production. Agriculture is a vast and necessary industry, and far beyond the scope of the few paragraphs written here. Agriculture is a tremendously broad term that encompasses the cultivation of food and fiber, medicinal plants, raising of animals, production of mushrooms and other fungi, and the many products coming from the soil to sustain human and animal life. Because of this diversity, it is unfair to peg western water use as inherently wasteful, as some critics of the industry do. While one might point to many areas of inefficient use, it is just as easy to highlight extremely efficient operations. At the heart of the efficiency question is invariably economic considerations. As long as we as a society continue to push water to be used in the highest and best use in any given area, agricultural water use practices will continue to improve.

As an industry, from the nineteenth to twentieth centuries, agricultural labor rapidly transitioned from mostly small independent farms to large corporatized operations. Although the initial spurt of agricultural development during the modern era in the West was by colonists—the "yeoman farmers" as heralded by Thomas Jefferson—most construction actually occurred through the active involvement of corporate and state capital, in what some have called "subsidized monopolization" by the state.[34] Today large corporate farms continue to receive

Figure 2.5. The United States has long used low-wage labor to perform its most difficult agricultural tasks. Here, an American of Japanese descent harvests lettuce in San Benito, California, in May 1942. When Farm Security Administration photographer Russell Lee made this image near the start of World War II, this man was awaiting relocation to an internment camp. Library of Congress, Farm Security Administration Photograph 8c32037. Photograph by Russell Lee. http://hdl.loc.gov/loc.pnp/fsa.8c32037. Public domain image courtesy of the Library of Congress.

huge subsidies from the federal government through direct payments, indirect support through infrastructure construction (including low-cost water), tax breaks, and other support. Small farms receive significantly less federal largesse. The agricultural workforce, which now constitutes only about 2 percent of the total US population, can be considered rural industrial, agricultural laborers. Much of this workforce find themselves managing industrial-scale equipment, crops, and animals within corporatized organizational structures. Only in those sites where industrial equipment has not replaced hand labor does one find organized troops

Figure 2.6. This is an oblique satellite photograph of an industrial scale feedlot near Grand View, Idaho. The feedlot, which is over three-quarters of a mile across, contains so many cattle that they appear as black specks. Water use in sites such as this is highly efficient. However, the high density of animals at sites like this can concentrate pollutants in runoff and groundwater. Image source: screen capture Google Earth.

of migrant laborers harvesting produce from the fields. Even farm owners find themselves cogs in the wheels of commodity-centered agribusiness. Small farms and ranches persist, but their owners find that they must specialize or excel in the production of certain farm products to remain competitive with large operations that produce for distant commodity markets. The exceptions to these trends seem to underscore the industrial shift in agricultural production: upmarket farm-to-table restaurants and farmers' markets glorify the small ag operations that persist as islands in a sea of industrial grain, produce, and livestock producers.[35]

ALERT

As climate change exacerbates extreme weather events, communities need reliable systems to warn them of potential flood events. Fortunately, just such a system exists. In storm and flood emergencies, the "Automated Local Evaluation in Real Time" network, known simply as ALERT, provides an early warning for high rainfall events that can cause severe flooding and damage. The network does this by bringing together a system of gages (mostly rain, but also streamflow and other information) and transmitting it to local authorities. The National Weather

Figure 2.7. An ALERT Station outside of Fredonia, AZ. The tube is open to the air on top, and contains a gage that measures the rate of rainfall. Technicians set the threshold for sending signals, which they tie to rainfall values that can trigger floods. The signal is relayed to a central location where technicians monitor the local array of ALERT stations. Author photo.

Service sponsors the ALERT system and shares the data with various state and local agencies. Each ALERT station is a series of linked gages and sensors. Data from each ALERT station is compiled and uploaded it to the internet to give real-time pictures of flood events across broad areas. Gages are generally powered by 12-volt batteries connected to small solar panels attached to the top or sides of the gage.

ALLUVIAL FAN

In the early twentieth century, geologists studying sedimentation began differentiating between delta sediment deposited into lakes or oceans and sediment deposited onto dry land. The latter, called *alluvial fans*, form at the foot of mountains. Closely associated with the desert regions in the American West, these wedge-shaped sedimentary deposits of stream alluvium and debris flows radiate down and out from canyons and onto valley floors. The classic fan shape arises from the stream jumping out of its channel and migrating to a different place on the slope, a process known as avulsion, during various storm events. Over time, multiple avulsion events take place, and the creek migrates back and forth to many sites along the growing wedge or fan of alluvium. From the air, the classic sediment wedge of an alluvial fan resembles an outspread hand fan. Near the canyon mouth, the sediment wedge is thickest, and the debris coarsest. The wedge thins and the sediment becomes finer-grained the further the stream gets from the canyon. Many streams forming alluvial fans are known as losing streams due to diminishing stream flow as water seeps into the sediment. Active alluvial fans have surfaces where sediment deposition is still occurring after storms. When deep erosion gullies cut alluvial fans, this is evidence that the fan is inactive. You can see alluvial fans across the West, but the best examples flank the mountain ranges of the Great Basin and Mojave Desert of Nevada, California, and Arizona. In this region, large faults have simultaneously raised the mountains and lowered the basins to create the topography conducive to alluvial fan formation. With this faulting, the mountains serve as the sediment source and basins as the place where the sediment gets deposited. Here, multiple alluvial fans typically coalesce into enormous sediment-filled basins.

These vast alluvial fan deposits comprise the repositories of huge groundwater supplies. For example, within the Basin and Range province of Utah, approximately 85 percent of the water withdrawn from wells comes from these alluvial deposits. Because of the arid climate, these alluvial aquifers recharge very slowly, making them nonrenewable resources unless artificially recharged.

Figure 2.8. Here is an oblique satellite image of a classic alluvial fan in the Badwater Basin area of Death Valley, California. The fan near its base is about 3,500 feet across, and over 450 feet high. Screen capture of image courtesy of Google Earth.

In Utah alone, these deposits contain an estimated 70 million acre-feet of water in the uppermost 100 feet of sediment.[36] By comparison, Lake Powell can store about 24.3 million acre-feet of water. Moreover, these alluvial deposits often support desert springs and cienegas. These springs commonly serve as the sole sources of water for wildlife and cattle and also support wetlands that are home to rare and endemic species.[37] Since alluvial fan deposits contain vast groundwater supplies, municipalities and agricultural interests have exploited their resources. As a case in point, the hotly contested (and now abandoned) plan by the Southern Nevada Water Authority to drill wells in Basin and Range alluvial deposits and pump it to the Las Vegas area demonstrates the economic and ecological value of these deposits.

ALLUVIUM

When you walk or drive across the vast valleys of Nevada's Basin and Range province, you're crossing massive accumulations of alluvial sediment. Deposited over tens of millions of years as mountains rose and eroded in response to

Figure 2.9. Here is exposed alluvium on the riverbank of the Green River in Dinosaur National Monument. Erosion, likely due to changes in river flow related to the construction of Flaming Gorge Reservoir upstream, has caused the Green River to cut down into its alluvial valley leaving behind a perched floodplain that now only rarely floods. Large Fremont cottonwoods, such as these, can only grow to maturity if the root systems can tap groundwater flowing through the alluvium. The sandbar in the foreground that the rafter is standing on is also alluvium but is part of the active river channel. Photo courtesy of the author.

Continental Drift, these valleys often contain thousands of feet of alluvium. So, here in one of the aridest environments of the American West, you see seemingly boundless amounts of sediment that moved to its resting place through the action of water.

Geologists call these accumulations of clay, silt, sand, and gravel deposited by rivers in a streambed, flood plain, delta, or at the base of mountains, *alluvium*. These deposits often hold significant reserves of sand, aggregate, precious metals, and sometimes gems. Aquifers within unconsolidated alluvium often contain critically essential aquifers. Arizona cities, such as Phoenix and Tucson, rely on alluvial aquifers for much of their municipal water supply. Alluvium provides a matrix for groundwater to accumulate, and water flowing back and forth between the aquifer and the stream influence a river's base flow and annual discharge. Alluvial deposits also provide the sediment substrate for vegetation growing across a river's riparian zone. The term itself comes to us from the Middle English word *alluviun*, meaning "matter deposited by flowing water," that, in turn, is from the Latin *alluvius*, meaning "washed against."

AQUEDUCT

Open the Oxford English Dictionary and you'll find that it describes an aqueduct as an artificial channel for conveying water, typically in the form of a bridge supported by tall columns across a valley. If this brings to mind the famous Roman aqueducts of Southern Europe, it should, as the term is indeed derived from the Latin word *aquaeductus*, which describes these very structures. While this is the classic definition for an aqueduct, the term's contemporary meaning has expanded to include the components of municipal water collection systems, including ditches, flumes, siphons, tunnels, and other structures that engineers link together to bring water to cities.

Perhaps the most famous aqueduct in the American West is the one built by the City of Los Angeles and completed in 1913 under the direction of William Mulholland. Taking its water from California's Owens River Valley, the Los Angeles Aqueduct winds its way through numerous tunnels, siphons, and open canals to supply Los Angeles with much of its water. To build the aqueduct,

Figure 2.10. The dedication of the Los Angeles Aqueduct in 1913 brought out thousands of people for the ceremony. Courtesy of Los Angeles Public Library.

Mulholland employed a small army of laborers who, at peak, numbered 3,900 men. On average, they received $2.50 per hour for their efforts. To this day, the aqueduct remains a critical component in the city's water supply.

AQUIFER

Hydrologists refer to subsurface water-bearing geologic formations as *aquifers*. A salient feature of aquifers is that they store and transmit water to wells, springs, or rivers. Most hydrologists restrict the use of the term to those water-bearing

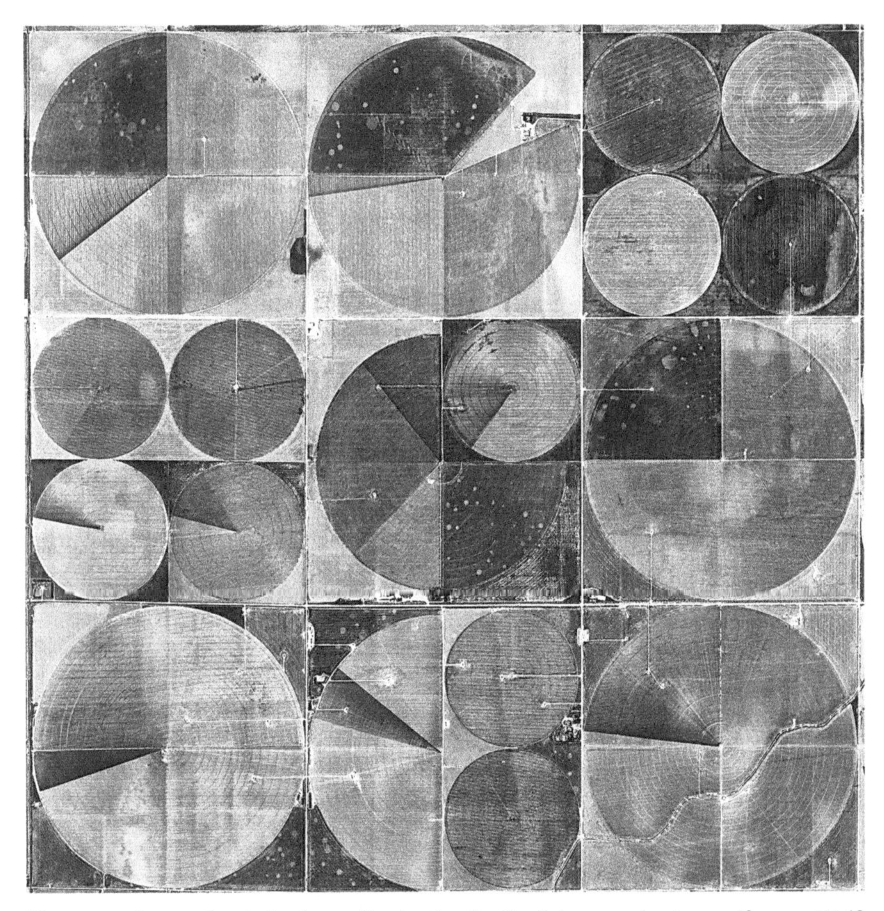

Figure 2.11. Center pivot irrigation wells piercing the Ogallala groundwater aquifer near Wolf, Kansas. Each large circle is about a mile across (or about three miles across for the image). Groundwater levels have declined by more than 100 feet in this area since wells were first dug in this area in the early 1990s. (https://www.climate.gov/news-features/featured-images/national-climate-assessment-great-plains'-ogallala-aquifer-drying-out). 2104 image screen capture from Google Earth.

formations capable of yielding water in sufficient quantity to provide a ready supply for people's needs. Conversely, hydrologists call formations that impede water flow or storage *aquicludes*. Another term that hydrologists often use is the *water table*. Hydrologists Thomas Dunne and Luna Leopold explain it this way: "Suppose that a well is drilled or dug into the ground. . . . At some depth, water will enter the hole and will attain a static level. The surface of the water in this well defines the *water table*."[38]

Water users have drilled many thousands of wells into aquifers to meet agricultural or municipal uses around the West. In many places, water users pump their groundwater excessively, to the extent that pumping exceeds the natural recharge of the aquifer, a situation known as "groundwater mining." Over-pumping of aquifers can lead to land subsidence and the diminishment of the aquifer's ability to store water. As groundwater mining proceeds, well production diminishes, leading water users to either deepen wells or drill more wells to maintain production in an ultimately losing battle against the decline in aquifer yield.

ARCHITECTURE AND MODERNITY

During the early twentieth century, rapid changes in all areas of life led people to believe that a new, modern world had emerged. Modernity and modernization became symbols of progress. Although vaguely defined, modernity was recognizable by its opposition to works commissioned by wealthy merchants and the church in prior centuries. Beginning in the nineteenth century and continuing into the twentieth, many artists started to make art based on their personal experiences and topics that they chose.[39] In art and culture, modernity expressed itself in a myriad of new forms: consider the cubism of Picasso, Art Deco, streamlined trains, Jazz Age music. Skyscrapers and other architectural marvels were explicitly modern in their appearance. Likewise, the engineering edifices of the emerging American Century would also shout modernity.

Perhaps nowhere would an engineering achievement express modernity so clearly as the greatest of them all: Hoover Dam. It became inextricably and inevitably linked to twentieth-century America. Later dams abandoned all artifice in the name of frugality and efficiency, and in doing so, they too would become distinctly modern. But with the Hoover Dam, although efficiency was paramount, the structure itself was deliberately designed to unmistakably express the sleek lines of Art Deco and incorporate artistic elements that the

Figure 2.12. Upstream face of Hoover Dam in 1935 or 1936. The art deco façade of the dam places this structure at the center of twentieth-century modernity. Bureau of Reclamation digital collection. Public domain image.

human eye would see in the smallest of details such as floor tiles and elevator doors. In doing so, the Hoover Dam became one of the instantly recognizable icons of modernism.

When engineers produced the first designs for Hoover Dam, it was clear that an architect was needed to balance the structure. The US Bureau of Reclamation awarded architect Gordon B. Kaufmann of Los Angeles the project, and they implemented most of his suggestions. Overall, Kaufmann simplified, ordered, and streamlined the dam's various elements and reworked the powerhouse into a modernist, stripped-classical style. Even so, additional ornamentation was believed necessary. The US Bureau of Reclamation placed surrealistic thirty-foot-tall winged creatures at the site and a star map on the crest of the dam that depicts the celestial alignment at the time of the Hoover Dam's dedication.

Hoover Dam represented a high point in integrating aesthetic considerations in the design of massive dams. But although no major American dam that followed had the extensive ornamentation of Hoover Dam, they nevertheless represent a unique aspect of American modernism.[40]

ARROYO

Another Spanish term that entered the English language via the West is *arroyo*—an erosional gully or channel found in arid or semiarid areas. When Spanish-speaking people use the term, they understand arroyos to flow continuously or intermittently. This is probably the most significant distinction with its use in English. For English-speaking people, arroyos characteristically have intermittent streamflow. Their shape is rectangular to steeply sided and trapezoidal to "V" shaped in cross-section. As erosional features, arroyos often cut into unconsolidated alluvial sand, silt, gravel, and soils. The channels themselves are usually at a minimum of several feet deep, often much deeper. In the Middle East and North Africa, these channels are locally known as *wadi*.[41]

River terraces and canyons across the West demonstrate the existence of arroyo formation for millennia. However, arroyos have expanded in extent over the last several centuries with the introduction of cattle and sheep from Europe. The introduction of livestock roughly correlates with the onset of the Little Ice Age (approximately 1550 to 1850 AD). Wetter conditions allowed Native and Hispanic peoples of the Southwest to establish and maintain large herds of livestock through that period. Once warmer and drier conditions set in sometime after 1850, the large herd sizes were no longer sustainable given the reduced forage. This, it seems, led to extensive gullying throughout the region. For more than a century, researchers have recognized that intensive grazing and the trampling of grasses have dramatically increased erosion in loosely consolidated alluvial material. However, the climatic push to this process has only been more recently understood. Once cattle and sheep trails were established and covering grasses were consumed in the watersheds, the underlying soil and alluvium became more susceptible to erosion. With summer thunderstorms and snowmelt runoff, formerly stable valley bottoms began eroding. Eventually, as T. Mitchell Prudden put it, a "great winding channel has been cut in the soil, which deepens and widens with every flood, so that today in countless instances along the erstwhile level grass-clad bottoms, over whose surfaces large volumes of storm-water once ran off harmlessly, are those ragged chasms called arroyos, which often reach to the rock bottom of the valley and from whose sides hugh [*sic*] masses of earth crack and fall to be swept along and away by the next flood."[42]

Consequences of arroyo development include declining groundwater tables, changes in vegetation patterns in valley bottoms, removal of water from historic floodplains, and topsoil loss. As the groundwater tables drop, soils in the valley bottom receive less moisture, and the vegetation shifts toward more arid

Figures 2.13 and 2.14. Here are two photographs of arroyos in New Mexico. The first at Chaco Canyon shows the former valley surface that was farmed by Ancestral Puebloans. Sheep and cattle introductions after the Spanish occupied New Mexico and the acquisition of these animals by various Indigenous tribes likely led to the initiation of arroyo cutting. At bottom is an arroyo at Albuquerque, NM, that was paved in concrete and flanked by Interstate 40. Water in this arroyo is conveyed towards the Rio Grande and does little to recharge the local aquifers. Photos courtesy of the author.

species. Areas that were once arable and farmed by Indigenous people in pre-contact times now appear unsuitable for farming because deep erosional cuts have formed in the valley bottoms.

ARTESIAN WELL

Around 1126, Carthusian monks in the Artois region of northern France began digging wells. To their surprise (and no doubt, great delight), water from their wells flowed to the surface without the need for any pumping. We remain in debt to these monks for bringing us the term "artesian" that remains in use to this day for any well water that flows up to the surface under its own pressure. In the American West, various attempts were made to develop artesian wells during the nineteenth century. Perhaps most famously, in 1883, a Mr. R. R. McCormick was boring for coal near the center of Denver, Colorado, when he struck water, forcing the abandonment of his effort. This caused some commotion, in

Figure 2.15. The horse-drawn carts located next to this artesian well in Kern County, CA, give a sense of the magnitude of the water coming from the confined aquifer below. Carleton E. Watkins, photographer. (Library of Congress Call Number: LOT 3379; Reproduction Number: LC-USZ62-53895, Public domain image.)

particular because the water was of higher quality than the nearby South Platte River, and because one could drill right on one's own property and get water. Within short order, people began digging many wells to supply all manner of uses. Of course, like all good things, overuse leads to diminishment, and within a few years, one by one, these wells stopped flowing.[43]

The science behind these flowing wells is that groundwater aquifers are sometimes bounded by tightly compacted low permeability formations. When these bounding formations, or aquitards, are extensive enough, aquifers lying below them can become confined with the result that water in them accumulates considerable pressure. After well drillers place a tightly cased well into one of these aquifers, the water pressure may push the water far above the surface of the aquifer. With sufficient water pressure, the water will rise above the ground surface, and the resulting *artesian well* will flow without the aid of pumping. Of course, if enough artesian wells are then drilled into the aquifer, the pressure in the aquifer will eventually diminish, and the wells will stop flowing, and one will need to start pumping to get at the water.[44]

ART OF EMPIRE

Water permeates the broader culture through art, literature, and cinema. Our cultural engagements with water help shape people's opinions and understandings of water and the environment. In particular, greater awareness of rivers and lakes and what they represent to people have emerged through dialogues surrounding landscape-scale artworks such as Christo and Jeanne-Claude's proposed installation *Over the River* and Robert Smithson's *Spiral Jetty* from 1970. Elsewhere, the monumental scale of waterworks evokes in the mind images of the greatness of the American Empire: here, desert dams authorized by American presidents elicit comparisons to the Pharaohs of Egypt who ordered the pyramids built. It is no surprise then that Oskar J. W. Hansen, who designed the towering statues adorning the Hoover Dam, took his inspiration from the Egyptians and sought to have future generations ask "what manner of men were these" who could build such structures? Hansen's statues bear deliberate but surrealistic resemblances to ancient Egyptian gods in ways that force the viewer to draw parallels between the great wonders of the ancient world and those wonders built by contemporary desert civilizations.

Figure 2.16. Oskar J. W. Hansen's towering Egyptian-inspired statues adorn the top of Hoover Dam and suggest an American Empire that conquered the arid west, just as the Pharaohs did millennia before along the Egyptian Nile. Image courtesy of the US Bureau of Reclamation, public domain image.

ASPINALL, WAYNE

As a United States Congressman from western Colorado, Wayne Aspinall (1896–1983) became one of the most influential politicians in the development of western water resources. Aspinall served in the US Army during both World War I and II. During World War II, Aspinall held the rank of captain. He participated in the D-Day landings at Normandy in 1944 and served with the US Eighth Army in its drive toward Germany. After the war, Aspinall entered politics and was elected

Figure 2.17. Congressman Wayne Aspinall deftly pulled political strings in Washington to position himself at the center of the major water policy decisions taking place during the mid-twentieth century. His influence helped shape the course of western development for decades to come. Image courtesy of the University of Wyoming, American Heritage Center, Floyd E. Dominy Papers, Accession Number 02129-1981-05-26, Box 8B

to represent Colorado's Fourth District in the US House of Representatives, serving from 1949 to 1973. Aspinall eventually became chairman of the House Interior and Insular Affairs Committee, where he supported dam construction and desert reclamation, including the Colorado River Storage Project. In this bill alone, Aspinall sponsored the proposed dam at Echo Park in Dinosaur National Monument; the Glen Canyon Dam, which created Lake Powell; plus the Flaming Gorge, Navajo, and Curecanti Reservoirs. For his support of these projects, Aspinall landed five other reclamation projects and three hydroelectric dams for his Fourth District. During his time in Congress, he was heavily involved in public land issues. Known as one of the most conservative members of Congress, this Colorado Democrat initially stalled the passage of the Wilderness Act that was being promoted by John F. Kennedy. However, after Kennedy's assassination he relented, as he did not want to be seen as insulting the public's empathy toward

Kennedy. Toward the end of his career, his conservatism and open hostility toward the environmental movement led to his being named to a 1972 "Dirty Dozen" list of biggest congressional enemies of the environment. Shortly after that, Aspinall lost a primary fight to a more liberal Democrat and fell from his perch at the top of the greatest dam-building spree in American history.[45]

AUGMENTATION

In the list of obscure technical water terms, augmentation surely ranks as one of the most elusive and arcane. Still, it is a powerful tool in the water manager's toolbox. For water wonks, augmentation refers to supplementing established water supplies. These additions could include drilling wells to supplement ditch rights at a farm, trans- or interbasin water transfers to a city's major reservoir or watershed, perhaps rainmaking to increase water to a watershed, or

Figure 2.18. Here, Alan Frank, right, and his son, Erik Frank, prepare to move siphon tubes to flood one of their cornfields near Gilcrest, CO. In 2007, the Franks had 12 wells shut down by the State of Colorado because they did not have sufficient augmentation water to offset the effect of their wells on senior water rights downstream. Frank's wells are in the Central Colorado Water Conservancy District Well Augmentation Subdistrict, which supplies their augmentation water. Still, due to the dry conditions, there was insufficient water to offset impacts to senior water. With his wells shut, Frank now grows less corn and more crops such as winter wheat and sunflowers that require less water. Photograph by Darin McGregor, Rocky Mountain News. Source: Image courtesy of the Denver Public Library, Special Collections http://digital.denverlibrary.org/cdm/singleitem/collection/p16079coll32/id/212913/rec/1.

desalinization to provide additional drinking water during a drought. Occasionally, but less frequently, hydrologists describe augmentation as a type of artificial recharge to extend the life of aquifers.

Elsewhere (and particularly in Colorado), augmentation is used in a specific way to describe a plan or program that allows junior water right holders to use water when facing a call from a senior right. For example, when a water user has a junior right, often from a well, and the owner of a downstream senior right calls for water, the local water commissioner may curtail or shut off the well to satisfy the senior right. However, if the well owner secures a source of supply (generally from a reservoir that can deliver water to the senior right), the well owner can then continue using his or her well. Water users call the alternative supply provided to the downstream right an augmentation source, and the program to deliver water an augmentation plan. When practiced carefully, this type of augmentation has evolved into an important water management technique that can allow junior water right holders to divert without harming senior water right holders who would otherwise shut them down. In some instances, however, pumping has effectively disconnected the aquifer from the river so that senior users experience harm. Often the water purchased for an augmentation supply was formerly put to lower value uses than either the senior right or the junior right that is then able to continue diverting. In prior appropriation systems, augmentation plans help maximize the value of water used within a watershed and allow junior water right owners to continue using water when they would otherwise get shut down.[46]

AVALANCHE

Avalanches mostly contain snow but often entrap ice, soil, rock fragments, and even trees as the rapidly moving mass accelerates down a slope under gravity. Although the term almost universally applies to the sudden movement of snow and ice in mountainous areas, geomorphologists also cite very poorly sorted deposits resulting from similar slope failures of destabilized soil and near-surface rock material as debris avalanches.[47] During summer thunderstorms, avalanche chutes often conduct debris flows down their course. This debris makes avalanche chutes one of the most geomorphically active water features in the American West.

Between 1950 and 2017, 1,056 people died in avalanches, making this one of the most dangerous natural hazards in America. Colorado and Alaska lead the nation in avalanche casualties. In most of these accidents, the avalanche fully buried its victims, with many dying either from suffocation or traumatic wounds. In Colorado alone, more people have died in avalanches than any other natural hazard. This sad

Figure 2.19. This avalanche, left, on March 7, 2018, covered Colorado Highway 91 about one mile south of Copper Mountain, CO. The same avalanche, right, buried this car on the highway. Fortunately, the passengers survived. Photograph courtesy of the Colorado Department of Transportation and the Summit County Sheriff's Department. Public domain photographs.

pattern will likely continue since accidents are generally tied to the growth of popular sports, including climbing, backcountry skiing, and snowmobiling.[48]

AVANYU

Water, for the Indigenous people of the Southwest (and for that matter most other Indigenous people), is sacred. Water represents life and contains power. Since water is so central to the Indigenous identity, water symbols and signs permeate Indigenous mythology and religious practice. One Puebloan deity, in particular, stands out for its essential association with water. This Puebloan deity is the *Avanyu*, alternately known as the "water serpent" or "Horned Water Spirit of the Below World." The identity of Avanyu varies among the tribes, but they generally consider him a water guardian. The Hopis call the water serpent *Paluukong*, and the Zuni call him *Kolowisi*. Likewise, the Huichol of Northern Mexico revere a water serpent. According to the anthropologist M. Jane Young, the "Zunis believe that all bodies of water—springs, lakes, ponds, rivers, and so on—are connected by a system of underground waterways to the surrounding oceans; Kolowisi travels through these underground waters to reach the various bodies of water that serve as outlets to the surface of the earth."[49] Within the ancient Pueblo world, the reasons for rain and drought were ascribed to supernatural causes. To pay homage to a deity such as Avanyu was a prudent way to interrupt cycles of drought and assure themselves of a reliable water supply.

Figure 2.20. Several spectacular depictions of Avanyu, the Puebloan water guardian deity, gather together on this boulder in a remote canyon in northern New Mexico. Photo courtesy of the author.

Prior to colonization, Indigenous people possessed sophisticated engineering knowledge that allowed them to construct diversion structures and ditches to move water, and build retention structures and reservoirs to store water. Avanyu demonstrates an additional form of indigenous knowledge related to groundwater and the hydrologic cycle. For these people, the control of water is both a spiritual and practical matter.

The former governor of Hémes (Jemez) Pueblo, Paul Tosa, has written that Avanyu images signify water: "In some cases, it marks the location of a water source. In others, it is a prayer related to our songs and dances as we ask the clouds to come and bless us with their moisture."[50] When depicted on petroglyphs or pictographs, Avanyu's curved lines simultaneously suggest the zigzag of lighting, the ripples of flowing water, or the meanders of a river. Lightning itself is mighty, always associated with clouds, thunder, and rain. Worldwide, people see serpents as a powerful metaphor for fertility. Since *Avanyu* presides over water, his importance for nourishing plants and animals is unmistakable. Puebloans generally depict Avanyu as a horned serpent. In particular, it is Avanyu's horns that evoke the spiritual power of this entity. Avanyu appears as part of a continent-wide homage to snake beings among Indigenous people. Specifically, the horns suggest links to the Quetzalcoatl religion of the Aztecs and Post-Classic Maya. Additionally, snake symbols were consequential to the Mississippian peoples (Cahokian Culture). Snake symbols persist among the Indigenous peoples of North and Central American and generally reference water, power, and fertility.

BANK

If you are a fisherman, river runner, hiker, or just about anybody who gets outside, you've probably scrambled up a riverbank at one time or another. And if you are interested in streams, it is worth considering riverbanks for a moment as something more than an obstacle. Geomorphologists describe a stream bank as a "sloping margin of a natural, stream-formed, alluvial channel that confines discharge during non-flood flow." For river runners and earth scientists, a right or left bank is determined by looking downstream. Stream banks are created by the balance between the erosion of sediment by flowing water, the deposition of sediment, and the cohesion provided by vegetation. The size of the sediment grains that are present, including their size distribution, influences the channel

Figure 2.21. In this photograph of New Mexico's Chama River, the water is just at the point of bank full discharge. Photo courtesy of the author.

and bank structure. Hydrologists have spent considerable time working out the details of how stream banks form in connection to a stream's discharge.[51] A major focus of their research revolves around answering the thorny question, What discharge sets the size and the shape of a river channel? A related concept is bank full discharge that, as the term implies, is water flowing at the top margin of the bank. Once the flow exceeds the bank full discharge, the water moves out onto the floodplain. Determining bank full discharge is economically significant for planning purposes since once water spills out of the bank, a river is considered as flooding. As water overtops the riverbank, water spreads out on the flood plain and loses velocity. As the flood event recedes, sediment gets deposited on the floodplain. Often the most abundant quantities of sediment settle out on the bank itself, and a natural rise or levee may result.[52]

One can often see both erosional and depositional banks in the same reach of river. The channel floor is a dynamic interface between water and sediment that can change from erosion to deposition in a matter of feet or with small variations of discharge. Stream banks come in a variety of shapes, from nearly vertical to very gentle slopes. A steeply sided to vertical bank generally indicates a site of erosion, and a gently sloping bank is often the site of deposition or moderate erosion. Vertical banks erode by undercutting that leads to sediment blocks collapsing into the river. Saturated banks are particularly prone to failure because soil water reduces the effective strength of unconsolidated materials. Densely vegetated banks containing complex root systems are far more stable than similar banks with fewer roots. Similarly, banks dominated by bedrock or boulders are more durable than banks consisting of finer-grained material.

BAR

Years ago, I had the opportunity to make a canoe trip along the famous Missouri Breaks of central Montana. In planning for the trip, I had heard that the entire run was flat water. But, when I examined the USGS topographic maps of the area (in the days before the internet), I became concerned about the many named rapids shown on the Missouri River channel. It turns out that we had nothing to fear, as these were actually submerged gravel bars that posed no threat to canoes. Instead, they were a legacy of the steamboat era that provided a warning to riverboat captains that their much larger craft might get lodged on these river deposits.

So, it really does not matter whether you have a canoe or a steamboat, a river bar is something that you will eventually encounter. Simply put, a *bar* is an accumulation of sediment within a river channel.[53] The sediment typically varies

Figure 2.22. This is a point bar on the Yampa River in Dinosaur National Monument. Photo courtesy of the author.

in size from coarse sand through cobbles. This material is generally deposited during the recession of a high flow event and becomes exposed during periods of low flow. The seeming irregularity of bar distribution may create hazards for river craft. Depending on the depth of the water, or river stage, one may float harmlessly over a bar or gets stuck and have to deal with the consequences. In perennial streams, the upper surface of bars often emerges from the water at about 40 percent of the average high-water flow.[54] Sediment bars belong to the large variety of streambed forms that are found within river channels. Nineteenth-century gold miners targeted river bars for the gold deposits they possessed. More recently, aggregate miners have exploited river bars for the sand and gravel they contain. And, when buried, the coarser river bar sediments often make excellent aquifers. River bars, it seems, have importance beyond their presenting hazards to river boat captains.

BASE FLOW

I'm often amazed that rivers can sustain flow year-round. Many rivers get quite low during dry seasons, sometimes nearly dry during droughts, but nevertheless maintain some flow, even if tenuous. This is because there is sufficient groundwater within the river's alluvium and surrounding watershed to provide a base flow. For hydrologists, *base flows* are the sustained flows that occur within a stream in the absence of direct runoff. Groundwater discharging to the stream

Figure 2.23. The Rio Grande River near Del Norte, Colorado, in early morning light showing base flow conditions on a winter day. Note how the ice, which has formed during base flow conditions, is located at a river level that is several feet below the adjacent river bank. Photo courtesy of the author.

channel throughout the watershed, including the river's alluvium, is the source of this water. The term does not distinguish whether the water in the stream comes from natural sources or is due to human modifications to a watershed's hydrology. In areas that primarily derive water from snowmelt, base flows are typically observed from the late summer through early spring. In monsoonal areas, or where thunderstorms provide the majority of runoff, base flows generally occur in between storms or storm seasons. These flows are not necessarily constant, but because they are derived from groundwater their fluctuations are modest. When looking at a stream in the fall or winter, the base flow height is well below that of the stream bank.[55]

BATHTUB RING

Take a close look at most western reservoirs and you will notice a discolored band of land devoid of vegetation lying just above the water level. These bands, or *bathtub rings*, arise from annual fluctuations as managers store and release water. They also occur from longer-term water level declines as watersheds dry

Figure 2.24. An aerial view at Lake Oroville, CA, shows the low water bathtub ring along the South Fork of the Feather River on January 16, 2014. Here, the light color band results from tree removal and the decomposition of vegetable matter that drowned when the reservoir was at a higher level. Photograph by Paul Hames, courtesy of the California Department of Water Resources. Public domain image.

due to global warming and climate change. Sometimes this band is black due to algae, and other times it may be white due to the precipitation of salts or the decomposition of vegetation. Elsewhere, the band arises from the settling of fine sediment when the water level is high and then later is exposed when the reservoir level drops. Furthermore, the seasonal drowning or drying of plants that require a more constant water supply leads to scant vegetation growth along the area of the ring. It's often a combination of factors that gives rise to the banding that you observe in many reservoirs. For reservoirs such as Lake Powell, the large bathtub ring it possesses has become a metaphor for desertification as streamflow in the Colorado River Basin has diminished.[56]

BEAVER

The North American beaver (*Castor canadensis*) is renowned for the dams and lodges it builds along rivers to pond water as protection against predators. It is also North America's largest rodent. Before its active hunting and trapping, many mountain rivers in the West had pool and riffle channel systems

Figure 2.25. Beaver on the bank of the Green River, Desolation Canyon, Utah. Photo by Mike Cecot-Schearer. Used with permission.

controlled by beaver dams. Habitat created by beavers influenced the aquatic species and riparian flora present along these rivers. As such, it was a keystone species along the West's streams. It's thought that the population once ranged from 60 to 90 million animals but hunting and trapping decimated the population in North America by the mid-nineteenth century. Once trappers removed the beavers, their dams began to fail, and the geomorphology of the streams shifted to eroded free-flowing channels. With the collapse of the fur market, beaver populations have steadily rebounded, and biologists now believe ten to fifteen million beavers live in western rivers.[57]

To get a sense of the importance of the beaver to American history, the next time you find yourself in New York City, take the subway to the Astor Place station. Here, you will find a peculiar nod to western expansion in the form of wall tiles imprinted with beaver images. Perhaps this is a bit tangential to western water's story, but these tiles commemorate how global commodity markets dramatically impacted western rivers in the nineteenth century. As some readers may recall, John Jacob Astor made his fortune by monopolizing the western fur trade, particularly the pelts of beavers, which made him the wealthiest man in early nineteenth-century America. Astor and his contemporaries extracted vast profits at the expense of western riparian ecosystems and the destruction of the Indigenous cultures that stood between the white man and his beaver pelts. Western riparian wealth propelled New York into the ranks of the world's great cities. To me, these tiles at Astor Place symbolize the rise of an American empire bent on colonizing the West and John Jacob Astor who stood at the forefront of this empire during the early nineteenth century.[58]

From the opening of the West to Euro-American exploration and exploitation to its role as a driver of geomorphic change, few—if any—species, have had a more significant impact or central role in the West's history, economic development, conflicts, and culture than the beaver. From colonial times through the 1840s, the beaver's lustrous pelt made the animal a coveted commodity that led to colonization, intrigue, competition, and war among European empires, Indigenous nations, and capitalists.

BELL MOUTH SPILLWAY

Spillways, the emergency relief valves that help prevent catastrophic dam failures during floods, take many forms based on the specific needs that engineers identify at a site. The most peculiar, known as a bell mouth spillway (sometimes

Figure 2.26. The bell mouth spillway at the Hungry Horse Dam, Montana, in which water cascades over the rim and drops 490 feet, is the tallest in the world. Photo taken in the early 1950s. Photographer unknown. Image courtesy of the Bureau of Reclamation. https://www.nps.gov/articles/montana-hungry-horse-dam.htm. Public domain image.

called a *morning-glory gate* or *glory hole* in old reports), is essentially a massive vertical drain within a dam. Due to their expense, engineers construct bell mouth spillways only when other options for conducting water away from a dam are limited. The elevation of the lip of the spillway establishes the maximum water height that a reservoir may contain. When the dam is full, and the bell mouth spillway is active, the structure presents a remarkable spectacle.

BENEFICIAL USE

When one talks about water in the West, there is a seeming multitude of legal terms that are invariably encountered. Among the most important of these is *beneficial use*, a legal phrase that refers to appropriations in which water is utilized in a reasonably efficient manner. Initially, the term came about to rein in clearly wasteful water uses. For instance, an irrigator who applies far more water than his or her crop needs is effectively depriving other users of the water that they need for their operations. In other words, water used wastefully is being used in a nonbeneficial manner. This notion of beneficial use came about in the age of hydraulic mining, particularly in circumstances where various placer miners were seen running far more water in their flumes than they actually needed to the effect that they were denying water to other

Figure 2.27. The beneficial use of water is illustrated by the careful flood irrigation of potato plants near Merrill, OR. Water comes from the US Bureau of Reclamation's Klamath Project. Here a farmer is using 2-inch plastic siphons to introduce water to well-tended rows. Courtesy of the US Bureau of Reclamation. J. E. Fluharty, June 1946. KP-1200-R2.Public domain image.

miners. With many eyes trained on each other's water use, mining district organizers instigated rules to enforce efficiency so that more miners could share scarce water supplies. These ideas were soon translated to the new farming districts that were popping up around the West. Eventually, the beneficial use concept evolved to quantify the actual amounts of water necessary for various uses within water right applications.

Logical as this idea may seem, the concept has generated significant controversy, as one person's notion of "beneficial" may be construed by another as wasteful. For example, the idea initially did not include environmental uses of water, such as instream flows for fish or recreational flows for boating. However, proponents of these uses eventually convinced legislatures to recognize their importance. With enabling legislation, environmental applications for water became established. The result of all this is that many people stand at the fountain of water use, with each one contending that their use is efficient while their neighbors are wasteful.[59]

BOATING HAZARDS

Across the West, tens of thousands of dams, diversions, bridge abutments, low bridges, weirs, barbed wire fences, and other human-made structures interrupt the flow of water. Many of these structures present hazards—often potentially

Figure 2.28. It is somewhat rare to see riverbank signs announcing approaching hazards. This sign, on Colorado's South Platte River in north Denver, warns boaters of a diversion dam that could prove fatal if one failed to exit the river ahead of the drop. Photo courtesy of the author.

deadly—to boaters or other recreationists who float the streams. Unfortunately, there are few if any regulatory checks in place when structures are built or rebuilt to ensure that they present as little danger as possible to river users. Compounding matters is that boaters who take out and walk around such structures can get cited for trespassing if they cross private property to stay out of harm's way. Many boating hazards, such as low head dams, have no warning signs posted upstream from the structures to alert recreationists of the nearby danger. Particularly dangerous are those dams that cross the stream in a single straight line. In this case, a recirculating hydraulic wave often forms that can trap a person or boat underwater and lead to tragic results.

BOTTLED WATER

All of us encounter or use bottled water wherever we go around the world. Its convenience is undeniable, but there's a dark side to the story. Bottled water is not unique to the American West. Still, it merits inclusion here for the simple reason that at every western waterway you visit, you will invariably find decomposing plastic water bottles on the shore or in the weeds along the bank. Beyond the obvious pollution from the plastic bottles themselves, it is worth considering the extreme cost of corporatized water and the vessels used to deliver it. In my hometown, for example, a homeowner currently pays just 0.0034 dollars per gallon for water coming from her tap. Drive down the road to Costco and pick up a case of bottled water, she'll pay $7.68 per gallon or about 2,259 times more than at home. Wow! Go to the local corner store or gas station, and she'll pay considerably more. Of course, these costs vary widely across the country, but I think you get the point.

Many people believe the urban folklore that bottled water is purer than what you get from the tap, but for the most part, it is simply not true. Water quality laws require municipalities to perform frequent testing and water quality monitoring, and they are held to high standards of purity for water entering municipal pipes. And when they do find a problem, they are also required by law to report it to their consumers. That is not to say that problems don't occur. Tragic situations do occasionally arise and often become sensational news, news that bottled water providers happily capitalize on. The lead poisoning debacle in Flint, Michigan, is a sad case in point. But Flint is an exception that proves the rule. Plus, bottled water providers happily support the myth that their product is something special. Just look at the labels suggesting that their water comes from remote or pristine springs. In fact, upwards of 45 percent of all bottled water is repurposed municipal tap water. Moreover, studies of bottled water

Figure 2.29. Water bottles litter the shores of every western waterway. Here are a few that clutter the bank of Boulder Creek, near Boulder, CO. Photo courtesy of the author.

versus tap water demonstrate that bottled water is substantially worse for the environment.[60] Plus, companies are good at spinning yarns about their product. One Idaho bottled water company, for example, boasts that their water comes from the ground and is in the order of eleven thousand years old. Hydrologists, like myself, interpret this to mean that the company is pumping nonrenewable groundwater and marketing the depletion of their aquifer as a benefit. And, a Denver bottled water company gets its water from a commercial well in an industrial part of town. Even broadcasters renowned for their support of corporate interests agree that, for the most part, American tap water is safe. Fox News, for example, reports that "Water treatment plants across America do an incredible job of storing, cleaning and distributing water to our homes."[61] Lastly, consider the fossil fuels used to manufacture all those plastic bottles, plus the fuel and transport costs to get it to the stores. Not to mention that once reaching consumers, only a fraction of those bottles ever make it to the recycler; many simply get landfilled or worse wind up on a stream bank.

BOTTOMLAND

One hears the word "bottomland" quite often when discussing rivers, and while it is pretty clear that it refers to the lowest part of a river valley, it remains instead a loosely defined term. Geomorphologists have contemplated it the most and consider the bottomlands as the area of river valley underlain by alluvial sediments that have been transported and deposited by the stream flowing through the valley.[62] What is implied in this is that the bottomlands may include the channel bed and one or more river terraces. I generally agree with this characterization but would add that bottomlands include the trees, shrubs, and other vegetation supported by the stream that grow on the alluvium. I generally exclude dry terraces that only contain xeric flora as these areas of the river valley are likely outside the active stream corridor.

BROWER, DAVID

> "Dave Brower, sanctimonious bastard, was guilty of misleading the public with his full-page ads that made people think we're going to flood out the Grand Canyon from rim to rim, Grand Canyon National Park."—Floyd Dominy[63]

David Brower once remarked that he had "never met a tree that he didn't want to spike."[64] Brower was, of course, referring to the radical action by some

Figure 2.30. The Teton River as it looked in November 1971 before it was dammed by the ill-fated Teton Dam. The photo was taken at the future dam site and clearly shows the bottomlands that were inundated when the dam was constructed. In this area, the densely vegetated bottomlands are well defined because the canyon slopes steeply away from the river. The area encompassing the bottomlands includes the river itself and the forested area located between the meanders of the river. Photo Courtesy US Bureau of Reclamation. Cropped from original. Public domain image.

Figure 2.31. David Brower chatting with President Lyndon Johnson at the signing ceremony for the Wilderness Act on September 3, 1964. Brower is presenting Eliot Porter's book of photographs, *In Wildness Is the Preservation of the World*, to the president as others look on. Photograph by Cecil Stoughton. Cropped for clarity. Public domain image courtesy of the LBJ Presidential Library.

environmentalists of hammering nails into old-growth trees to make it dangerous and unprofitable for logging companies to cut. Such was David Brower: a scrappy soul hell-bent on using any method necessary—from direct action to political suasion—to protect the wildlands that he saw dwindling and ruthlessly exploited in America. So, it is a tad odd, but not surprising, that Brower is often inextricably linked to his lifelong nemesis, Floyd Dominy, the seemingly all-powerful commissioner of the US Bureau of Reclamation during the 1950s and 1960s. For it was the competition between these two men that in many ways defined them, and that likely made each even more famous than had they never met. Their greatest struggles centered on Dominy's desire to construct gigantic dams in the Colorado River Basin and Brower's resistance to those plans as the Executive Director of the Sierra Club. Their first epic clash erupted over the Bureau's proposal to construct a dam at Echo Park in Dinosaur National Monument. Brower pulled out all stops to scuttle it, placing full-page advertisements in the *New York Times* and other papers that asked whether America would drown the Sistine Chapel to store water. Brower's lobbying of Congress led to the Sierra Club losing its nonprofit status. But his ploy worked; Congress eventually got cold feet, and the project was mothballed. Yet this was not before Brower was forced to trade the Echo Park Dam for another one on the Colorado River that would eventually flood Glen Canyon and create Lake Powell. But Brower had never visited Glen Canyon before completing his trade-off. When Brower eventually did visit Glen Canyon, he remarked, "This makes me feel pretty damn sick," adding, "Glen Canyon is the greatest loss of scenic resources anywhere. . . . Down under that water, some of the most beautiful scenery ever created is gone."[65] For the remainder of his life, Brower regretted his compromise. Nevertheless, Brower later fought Dominy's plans for dams in the Grand Canyon, and he used many of the same tactics that were successful at Echo Park to kill those projects as well.

Brower's manyfold accomplishments beyond his opposition to dams would have made him famous. As a world-class mountaineer, Brower made over seventy first ascents. Brower also became the first executive director of the Sierra Club (1952–1969). He later founded the Friends of the Earth and the Earth Island Institute and cofounded the League of Conservation Voters. Plus, Brower was a prolific writer and advocated for wildlands. His lobbying was instrumental in the passage of the Wilderness Act and the establishment of 10 national parks and seashores, including Point Reyes, the North Cascades, and the Redwoods.[66]

BUREAU OF RECLAMATION

> "The Bureau of Reclamation—by the 1960s, a dam building, canyon flooding juggernaut."—William deBuys

When Theodore Roosevelt signed the Reclamation Act of June 17, 1902, he created a United States Reclamation Service within the US Geological Survey. Its initial location here, and not as an independent agency, attests to the influence beyond the grave of John Wesley Powell, the former USGS director who had lobbied tirelessly for the creation of this bureau. Activities of the Reclamation Service were limited to the western states. Immediately upon its inception, the Reclamation Service began building water projects to support and subsidize western rural and urban areas. In 1907, the agency was separated from the USGS, and in 1923 it acquired its present name. In 1928, Congress authorized the Boulder Canyon Project that resulted in the construction of the Hoover

Figure 2.32. The US Bureau of Reclamation Montrose, CO, Administration and Dispatch Building in 1966. With the look and feel of a mid-twentieth century space operations center, this US BOR facility utilized a "high-speed electronic computer which is the 'brain' for automated control of water releases, power production, and power dispatching functions for the Colorado River Storage Project transmission grid." Courtesy of the US Bureau of Reclamation. Public domain image.

Dam. Reclamation had entered its heyday. Dam building became a central feature of the American Century. Through the Great Depression and during in the Post–World War II years up until the 1960s, the era of big dams resulted in the impoundment of a multitude of western rivers.

With the collapse of the Teton Dam in 1976, the shaken agency was forced to reassess its priorities. That and President Jimmy Carter's so-called "hit-list" of water projects to be cut for budgetary reasons caused the big dam era to sputter to a virtual halt. Today, the US Bureau of Reclamation oversees about 180 projects in seventeen states that irrigate over nine million acres. Its fifty-four power plants produce over thirty-six billion kilowatt-hours of electricity. Although it seldom seems to make headlines today, Reclamation remains one of the most important federal agencies affecting the West's environment and economy.[67]

BUY AND DRY

Ask any farmer in a rural community about water, and you will quickly learn that one of the significant hot-button issues in western water is the transfer of agricultural water to municipal or industrial uses. Commonly known as *AG dry up*, or *buy and dry*, these transfers are explicitly allowed under prior appropriation systems in which water can be bought and sold. Once a water right is sold, the new owner goes through a court or administrative proceeding to move or transfer the water to the new use. When the water is transferred off of agricultural lands—usually to cities—restrictions preventing further irrigation are generally placed to avoid the expansion of use (new use plus old use) so that junior water rights are not harmed. These restrictions on further irrigation mean that the lands are dried up, hence the origin of the term AG dry up.

AG dry up can have considerable economic and ecological consequences in the affected area. Since irrigated land is almost always far more valuable than non-irrigated land, farming districts experiencing many water transfers often see

Figure 2.33. In the 1960s and 1970s, the Front Range cities of Thornton and Aurora began buying ranches for their water rights along the upper reaches of the South Platte River in South Park, CO. By the early 1980s, Aurora alone had transferred over 20,000 acre-feet of water out of South Park. As a result of this practice of buy and dry, the cities dewatered tens of thousands of acres of irrigated land and then abandoned ditches that had been operating since the mid-nineteenth century. Shown here is one such abandoned ditch near Hartsel, CO. Photo courtesy of the author.

their local economies contract as the intensity and diversity of farming diminishes. In extreme cases, primary and secondary sources of employment decline. For example, when water is transferred out of a valley, local irrigation supply stores may close, causing ripple effects to other businesses in an area. Ditches with fewer irrigators mean that the remaining farmers bear the increased burden of managing the system. Environmental impacts of transfers and AG dry ups include expansion of weeds, less groundwater recharge, lowered base flows in nearby creeks and rivers, fewer flows to dilute waste, and impacts to riparian vegetation and wildlife, to name a few.[68]

CALIFORNIA STATE WATER PROJECT

It seems that Californians rarely fear thinking big and following through on their dreams. One of the biggest of these plans was the passage of the 1960 California Water Resources Development Bond Act, in which California voters authorized

Figure 2.34. The Ira J. Chrisman Wind Gap Pumping Plant on the California Aqueduct in March 1989. This is just one of the many structures that comprise the California State Water Project. Photo courtesy of the California Division of Water Resources.

$1.75 billion to construct a vast north-to-south network of canals, pumping plants, power plants, and reservoirs to serve water users across the state. The system boasts several records, including the Oroville Dam, the tallest dam in the United States that, at 770 feet high, is one of the primary sources of hydropower on the project. Another component, the Edmonson Pumping Plant, completed in 1973, moves water over the Tehachapi Mountains and has the highest lift per volume in the world, with a single lift of 1,926 vertical feet and a maximum flow of 4,480 cfs. The American Society of Civil Engineers considers the California State Water Project one of the outstanding engineering achievements of the twentieth century. Project construction ran from 1961 to 2003, with ongoing maintenance and repair today. On average, the project delivers 2.4 million acre-feet of water and generates about 6.67 billion kilowatt-hours of electricity. The project moves water from the Feather and Sacramento Rivers through the Bay Delta where various pumps deliver water into the California Aqueduct southward past Pyramid Lake and on to water users as far south as the Mexican border.[69]

CANALS

Canals are the workhorse waterways that help enable life in the West as we know it. Human-made waterways used to transport water or facilitate transportation are termed *canals*. The word is often used interchangeably with both *ditch* and

Figure 2.35. This is a segment of the Friant-Kern Canal shortly after its completion on October 21, 1946. Courtesy of the US BOR, National Archive and Record Administration FK-520-CV. Public domain image.

acequia. If there is a difference between these features, canals are generally larger structures than either ditches or acequias. Even smaller yet are laterals, which are branches of ditches that people use to transport water to fields. Also, in the United States, the word *acequia* is rarely used outside of southwestern Hispanic communities.[70] Furthermore, both ditch and acequia seldom, if ever, connotate a waterway that is used for transportation (it is the Erie Canal, not Erie Ditch after all). People construct canals for a wide variety of purposes, including delivering water to cities, farms, mines, and other industries.

CANYON

Ever since conquistadores from the Coronado expedition first came across the abyss that we now know as the Grand Canyon in 1540, canyons have held a central role in our understanding of western geography. Essentially, a *canyon* is a deep, narrow valley surrounded by steep sides or cliffs. Canyons always have an ephemeral or perennial stream or river flowing through them. Often the term "gorge" is used interchangeably with canyon, but only in cases where the terrain is very steep and narrow. Canyon comes from the Spanish word *cañon*, which means "tube" or "pipe." In the literature of the American West up until the 1880s, it was common to see the word *canyon* spelled in its original Spanish, *canõn*, or anglicized as *canon*. By the 1890s, the current spelling settled into widespread use.

Canyons are erosional features carved by rivers. Water and wind erosion, along with chemical and mechanical weathering, promote the downcutting, or "incision," of geologic formations with the result that a canyon may form. Geologists often point to regional tectonic uplift as a mechanism for canyon formation. When a river becomes ensconced within a deep ravine, the course of the river is dominated by the surrounding bedrock. In these instances, the river is considered entrenched.

Sometimes a deep narrow canyon forms as sediment from surrounding formations are flushed downstream. When it is particularly deep and narrow, it is called a *slot canyon*. It calls to mind famous images of Antelope Canyon in Northern Arizona or the slots in Utah's San Rafael Swell. Slot canyons can be just a few feet wide but hundreds of feet deep. When they flood, the torrent is swift, furious, and potentially deadly, posing great danger to hikers. It is common to see logs or other debris jammed at seemingly implausible heights dozens of feet above the canyon floor, giving mute testament to the magnitude of past floods.[71]

Figure 2.36. This is the remarkably narrow slot canyon of Muddy Creek in Utah's San Rafael Swell in 2011. Notice the boaters beneath the debris lodged in the narrows. This logjam was in place when the record flow of 9,400 cubic feet per second (cfs) passed through the nearby gage in September 1981, which of course indicates that flow even higher than that deposited this debris sometime before the gage was installed. Muddy Creek was flowing at about 200 cfs at the time this photograph was made. Photo courtesy of the author.

CASCADE

On October 31, 1805, Lewis and Clark and the Corps of Discovery reached the Cascades of the Columbia River, their final trial before exiting onto the flat waters that would take them to the Pacific Ocean. Here, Lewis and Clark encountered "the great Shute," a four-mile stretch of river where the river ran "with great velocity foming & boiling in a most horriable manner [*sic*]." They completed their difficult passage by alternately floating and portaging the obstacles that lie in this difficult stretch. This whitewater was their final test of an arduous fifty-five-mile reach of the Columbia River that began at a formidable mass of whitewater known as Celilo Falls and ended here at the Cascades. It ultimately proved to be the most dangerous and challenging part of the Columbia River network that the Corps of Discovery descended in the entire Pacific Northwest.[72]

Adventures such as the Lewis and Clark expedition illuminated the difficulty of navigating cascades in the American imagination. As we know from Lewis and Clark and explorers like them, cascades are steep violent rapids or smaller waterfalls that drop in stages. These highly scenic features give a river the appearance of a more-or-less continuous tumult or chaotically turbulent mass of water flowing down a rocky river channel.

Figure 2.37. Here is a cascade along Spring Creek, a tributary of the Taylor River, in Central CO. Photo courtesy of the author.

CASH REGISTER DAM

In the mid-twentieth century, efforts to harness the power of nature peaked. The political forces of the era saw rivers from a primarily utilitarian perspective. To water users, water waste meant water that escaped diversion. Likewise, water managers believed that storing water in dams was the highest form of conservation. It was during this time that water users began building huge dams containing giant hydropower plants so that they might "peddle hydropower" to cities and farms and thereby subsidize the construction costs of the projects. In the 1960s, these hydropower projects acquired the nickname *cash register dams* that we use to this day. The term first came to public attention in a *Life Magazine* editorial describing plans to bracket the Grand Canyon with dams so that Central Arizona Project beneficiaries might receive water.[73] One dam was eventually

Figure 2.38. Most American presidents in the early to mid-twentieth century supported reservoir projects as good government policy and saw them as pork that could help swing future elections for their party. Here, President John F. Kennedy speaks at a short ceremony on September 27, 1963, at the Salt Lake City municipal airport at which he pressed a key to start the first generator at Flaming Gorge Dam. Source: US Bureau of Reclamation photograph by Mel Davis. Public domain image.

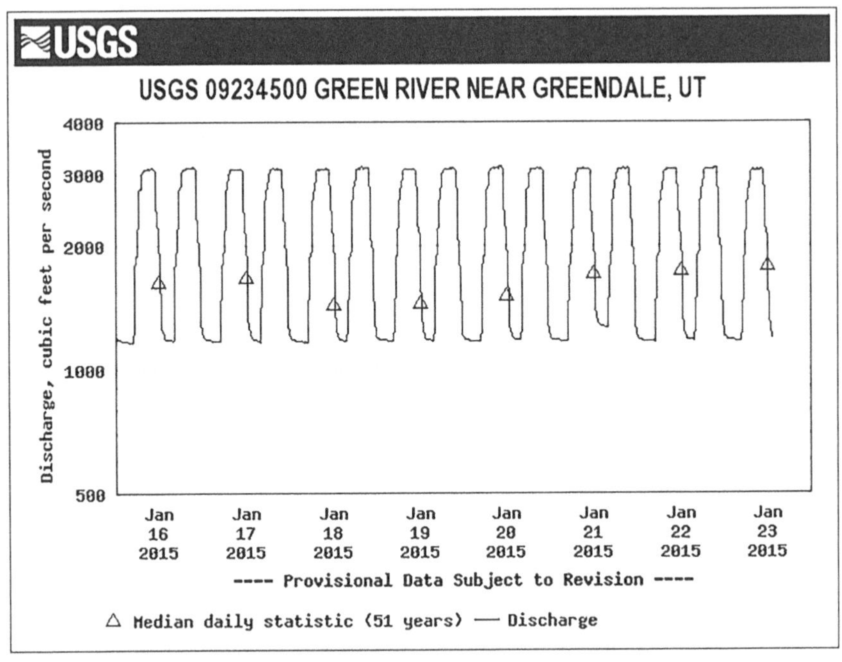

Figure 2.39. To illustrate how the cash register at Flaming Gorge works, consider this hydrograph of reservoir releases recorded on the Green River just below the dam. Notice the highly unnatural square-wave shape of the streamflow. Each day when electricity demand is high (and the price paid for electricity is highest), the Bureau of Reclamation opens the penstock valves. It then runs about 3,000 cfs of water through the turbines to generate electricity. Later in the day, when the electricity demand decreases, the Bureau of Reclamation cuts back the flow by more than half, to about 1200 cfs. On and on, this cycle repeats. Only occasionally will the Bureau of Reclamation adjust releases depending on electricity demand or due to changes in inflow from snowmelt or storms.

built just upstream from the Grand Canyon at Marble Canyon—Glen Canyon Dam—that impounded Lake Powell and flooded Glen Canyon to become the quintessential cash register dam. In time, the Bureau of Reclamation built many cash register dams around the Colorado River Basin, including Flaming Gorge, Navajo, and Morrow Point Dams, to name a few.

CENTRAL ARIZONA PROJECT

One of the most ambitious projects built by the Bureau of Reclamation, and certainly its most expensive, is the Central Arizona Project (CAP), a vital water system that makes much of Arizona habitable. Initially, the CAP was designed to irrigate nearly one million acres of agricultural land that were dependent on

Figure 2.40. Water diverted from the Parker Dam on the Colorado River flows into the canals of the Central Arizona Project. Photo courtesy of the author.

nonrenewable groundwater supplies. Once under construction, the purpose of CAP shifted from serving water to agriculture to a supplier of water for growing cities in this Sun Belt state. Its costs skyrocketed as the components multiplied: diversions, pumping stations, aqueducts, and tunnels were built to move water. CAP diverts water from Lake Havasu at Parker Dam and transports it across long stretches of the Sonoran Desert to Phoenix, Tucson, and as far as the San Xavier Indian Reservation south of Tucson. Operating CAP pumping stations requires so much energy that the Bureau of Reclamation considered plans such as damming the Grand Canyon for hydroelectricity or building nuclear power stations. When the public pushed back on those ideas, the Bureau of Reclamation retreated and became a participant in the coal-fired Navajo Generating Station.[74]

Construction of CAP entailed considerable controversy. California's politicians in particular argued that Arizona's previous hostility toward Colorado River development, including their refusal to ratify the Colorado River Compact for more than twenty years, disqualified Arizona from receiving such a massive federal subsidy. California raised a variety of issues, including the allocation of water in the Lower Colorado River Basin, to block CAP from authorization. Eventually, Arizona sued California in the US Supreme Court to resolve the issues. In 1963, the US Supreme Court ruled mostly in Arizona's favor, and this ruling reenergized attempts to authorize CAP.

Overall, it took nearly twenty-two years from the original introduction of legislation in 1945 to President Johnson's signing the bill authorizing CAP construction as part of the Colorado River Basin Project in 1968. Another five years transpired before groundbreaking ceremonies took place at Havasu Springs Resort in May 1973. Twelve more years passed until the first water deliveries to the nonIndian distribution system were made. Finally, CAP reached Phoenix in 1986, and in 1993 the US Bureau of Reclamation declared the Central Arizona Project substantially complete. After more than twenty years of construction, the federal subsidy that made life possible in much of Arizona was finished at a cost to US taxpayers of more than $4.4 billion. CAP's legacy is decidedly mixed. Economist W. M. Hanemann observed that on one hand, from "an engineering perspective CAP is, no doubt, a substantial success." On the other hand, "from an economic and financial perspective it must be judged something of a failure."[75] By the late 1990s, CAP provided an estimated 1.3 million acre-feet of water annually to Maricopa, Pima, and Pinal Counties in Arizona. More than eighty customers receive project water, 75 percent of which are municipal and industrial, 13 percent agricultural, and 12 percent to Indigenous tribes.[76]

CENTRAL VALLEY PROJECT

When you hear about Western water projects, people often use superlatives to describe these massive facilities. Of these, a few stand out beyond all others for their sheer size. This is undoubtedly the case with the largest and most expansive water project in America, if not the world, that is, California's Central Valley Project (CVP), a project that extends over four hundred miles from the Cascade Range in the north to the fertile Kern River Valley in the south. Initially, the CVP was designed to insulate California's Central Valley from drought, severe water shortages, and flooding. However, the system was quickly diversified to improve navigation on the Sacramento River, generate hydroelectricity, provide water for fish and wildlife habitat, offer recreational opportunities on its reservoirs, and deliver water to municipalities. Therefore, the CVP serves water across the breadth of California from cities in the Bay Area south to irrigators in Kern County. Operated by the US Bureau of Reclamation, the CVP provides water to six of the top ten agricultural counties in the country.

As early as 1873, engineers for the U.S. Army Corps of Engineers suggested the creation of a massive irrigation project in California's Central Valley. Nothing

Figure 2.41. California's Central Valley Project contains over 500 miles of major canals in its system. This includes the Delta-Mendota Canal, shown here in 1969, when the canal was flowing at 2,750 cfs. Image courtesy of the US Bureau of Reclamation. Photo by Wes W. Nell.

much came of that initial proposal, but within a few years of the passage of the Newlands Act of 1902, the Reclamation Service (now the US Bureau of Reclamation) began developing the project. However, years passed until a series of droughts and onset of the Great Depression motivated California and the federal government to build the CVP. California's CVP thus became one of Franklin Delano Roosevelt's most significant stimulus projects to help the United States climb out of the Great Depression. Work on dams and canals began in the late 1930s, and the project was substantially complete by the early 1970s.[77]

The CVP is truly an integrated system with twenty interlinked dams and reservoirs, eleven power plants, and about five hundred miles of major canals, plus conduits, tunnels, and related structures. The amount of water involved is also astonishing: annually, around 9 million acre-feet of water run through the system each year to deliver approximately 7 million acre-feet to all manner of users. Nearly 5 million acre-feet go to farms alone, providing water to 3 million acres, or approximately one-third of California's agricultural land. About 600,000 acre-feet go to municipal and industrial use and another 800,000 acre-feet per year to fish and wildlife.

How the CVP, a legacy of irrigation-era imagination and New Deal construction, will respond to the challenges facing California today remains an open question. The system faces acute stresses from population growth, changes to the agricultural economy, and climate change. Farmers receiving CVP water struggle with drainage issues and the buildup of toxic salts. Water quality issues in the Bay-Delta stemming from CVP diversions create ongoing tensions between the environmental community and water users. The decline in Chinook salmon runs in the Sacramento and San Joaquin Rivers remains a significant challenge. Proposals to increase storage and improve conveyance facilities must balance water user needs, environmental considerations, and provide for resilience in the face of potentially severe climate change. How well the CVP succeeds in facing these thorny issues will likely affect California's quality of life for decades to come.[78]

CHANNEL

Among the most dynamic environments on earth, a *channel* is a place where flowing water concentrates in a stream or canal. Channels follow a curvilinear pattern in response to local geomorphology. Factors affecting the shape of a channel include flow, sediment, vegetation, bedrock or alluvial substrate, and gradient, to name some of the significant factors. Erosion and sedimentation,

Figure 2.42. Range Creek (a tributary of the Green River in northeastern Utah) the morning after a flash flood eroded and deepened its channel. Photo courtesy of the author.

along with vegetation growth and decay and the ongoing impacts from humans and animals, make channels dynamic features. Even the channels of ditches are active due to the erosional potential of moving water. Human-made channels require ongoing upkeep or outright stabilization through the placement of concrete or piping lest they fill with sediment or erode their banks.

CHINATOWN

Roman Polanski's 1974 neo-noir mystery film, *Chinatown*, was inspired by Los Angeles's Owens Valley water grab. The movie, starring Jack Nicholson and Faye Dunaway, revolves around an illegal real estate scheme designed to bring water and riches to corrupt developers. Polanski chose this title, *Chinatown*, as a metaphor to suggest an opaque story in which key details are hidden from our view and illegal activity escapes investigation. The tone of the movie feels as if it was lifted from a Raymond Chandler novel. With a twisted plot and slowly leaking clues, the viewer is left guessing until the very end to understand sordid

Figure 2.43. A film still from Roman Polanski's classic 1974 noir film *Chinatown*.

family entanglements and the motivations for murder and corruption to accumulate the water, money, and power needed to fuel the growth of Los Angeles. The movie climaxes with (spoiler alert) a revelation of incest. Robert Towne, *Chinatown*'s screenwriter, put it this way: "The water scandal was the plot, essentially, and the subplot was the incest. That was the underbelly, and the two were intimately connected, literally and metaphorically: raping the future and raping the land."[79] Sorting this all out is J. J. "Jake" Gittes (Nicholson), a hard-boiled private investigator who winds up involved with the alluring Evelyn Mulwray (Dunaway), the wife of LA Water and Power's chief engineer. Mulwray's name is, of course, a thinly veiled allusion to William Mulholland, LA's actual twentieth-century water czar. Murder and intrigue follow, leaving Evelyn Mulwray dead and Gittes standing powerless in the face of a corrupt system. The film has rightfully received lasting acclaim, including its selection by the Library of Congress for preservation in the United States National Film Registry as a "culturally, historically or aesthetically significant" work.

CIENEGA

In the Southwest, hydrologists call perennial springs forming wetlands at the foot of a mountain, within a canyon, or on the low edge of a grassland where groundwater daylights *cienegas* (sometimes *cienagas*). In these sites, groundwater might come to the surface where shallow bedrock forces the flow upward. Unless pumped, groundwater levels tend to be stable and support the ecological diversity that is the hallmark of these systems. Like many other landscape words in the Southwest, cienega has Spanish origins.

When a cienega does not drain into a stream but instead discharges onto a low-lying area, it might form a salt pan, salt marsh, or playa lake. Cienegas often permanently saturate soils, giving rise to vigorous wetland fauna and flora. In the Southwest, cienegas represent less than 0.1 percent of the total land area yet harbor more substantial numbers of rare, threatened, and endangered species relative to any other landscape feature in the region. Abundant water also makes these sites appealing to wildlife and domestic animals as a source of forage. Because of ongoing grazing, many cienega soils are damaged and eroded.[80]

One of the more famous cienegas used to exist near the present site of Las Vegas, Nevada. Before settlement as a city, *las vegas* was a term used by Spanish travelers to describe the lush meadows that were supported by groundwater and springs at the site. In a genuine sense, this cienega was a critical transit stop for travelers crossing the Mojave Desert. Likewise, other lesser-known cienegas

Figure 2.44. Cienegas, such as this one in Northern New Mexico, provide water in an otherwise dry region. Nearby petroglyphs and ruins indicate that Indigenous people have used this site for thousands of years. Photo courtesy of the author.

served as essential stops for water and forage for desert travelers and often became first sites for settlement once the Spanish moved into the Southwest.

CLIMATE CHANGE

Climate change is a departure from the usual weather found in any given place.[81] A departure might mean that a particular locality is hotter or colder, or perhaps wetter or drier. Geologists have identified long periods of warming or cooling and cycles of drying or wetting throughout the geological record. In recent decades, however, much of the change around the world has been toward progressively warmer conditions. Since the late 1970s, atmospheric scientists have established a broad consensus that carbon dioxide emissions from human sources are warming the climate. More recently, the UN Intergovernmental Panel on Climate Change concluded that "It is unequivocal that human influence has warmed the atmosphere, ocean and land. Widespread and rapid changes in the atmosphere, ocean, cryosphere and biosphere have occurred." Moreover, they added, "The scale of recent changes across the climate system as a whole and the present state of many aspects of the climate system are unprecedented over many centuries to many thousands of years." Furthermore, "Human-induced climate change is already affecting many weather and climate extremes in every region across the globe."[82]

If heating trends continue, we may as well rename Glacier National Park to "Moraine National Park" to accurately describe the future landscape condition. Across the American West, the last vestiges of the Pleistocene Ice Ages are rapidly disappearing. As humans continue to pump carbon dioxide and other climate-warming chemicals into the atmosphere, the worldwide trend toward a hotter world will continue. The United Nations Intergovernmental Panel on Climate Change has compiled some of the best available global warming data. They report that since preindustrial times, human activities have caused approximately 1.0°C of global warming. At the current rate of carbon emissions, we can expect the warming to reach 1.5°C between 2030 and 2052.

However, the rate of carbon emissions is increasing. This increase means that the trajectory for warming is even higher. To limit warming to only 1.5°C, we need to require changes on an unprecedented worldwide scale that include deep emissions cuts in all economic sectors, deploy a range of technologies to limit emissions, effectuate human behavioral changes, and substantially increase investment in low carbon alternatives.[83]

The risk to the American West is vast. Climate modeling studies suggest that the Southwest will experience diminished rain and snowfall as the climate warms.

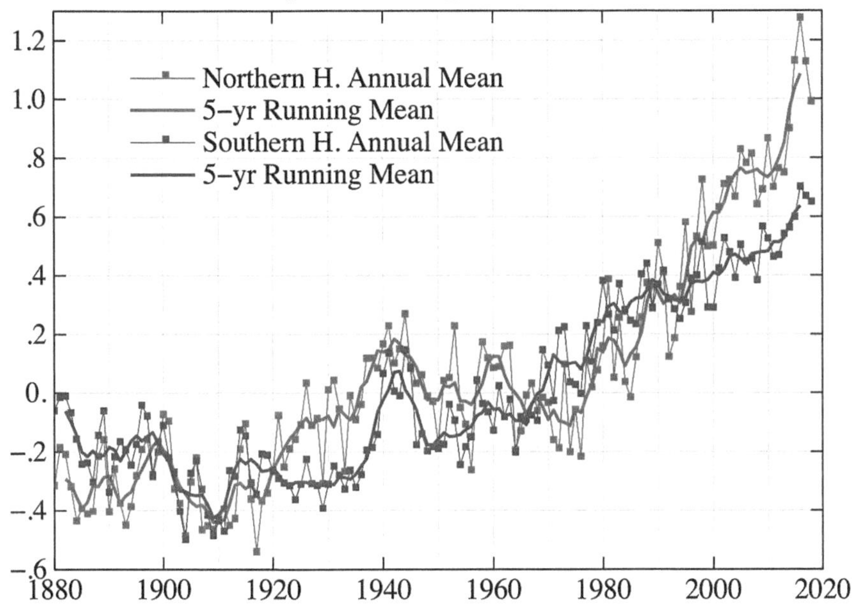

Figure 2.45. This chart from NASA tracks the steady rise of global temperatures since 1880. Source: http://data.giss.nasa.gov/gistemp/graphs_v3/Fig.A3.pdf. Public domain image.

Increased aridity will lead directly to reduced streamflow that will harm cities, farms, the environment, and industries alike. For example, researchers studying the Colorado River have estimated that flows have decreased by about 11 percent over the past century. They further estimate that if carbon emissions continue apace, Colorado River flows will decline an additional 19 to 31 percent by midcentury.[84] Already a desert, further aridification will make this region less habitable. Climate uncertainty can harm the region's economy as people and businesses rethink the wisdom of living in or relocating to places like the Southwest. Some researchers have already concluded that the increasing temperatures are causing more catastrophic flooding as well as wildfires that destroy forests, harm watersheds, and of course, cause terrible losses to anyone who lives in the fire's path.[85]

CLOSED BASIN PROJECT

Sometimes tectonic forces cause the earth's crust to move in ways that form deep isolated valleys. Some of these valleys, or *closed basins*, may have no natural outlet to the sea. Geographers use the technical term *endorheic basin* when

Figure 2.46 A canal and pumping equipment that is part of the US Bureau of Reclamation Closed Basin Project in Colorado's San Luis Valley. Photo courtesy of the author.

describing closed basins. America's most famous example is Death Valley. In these basins, precipitation or snowmelt can only leave by seepage or evaporation. A portion of Colorado's San Luis Valley contains a sizable closed drainage that intercepts water from the surrounding Sangre de Cristo and San Juan Mountains. Water draining to the middle of the basin collects in wetlands and soaks into the local aquifer or evaporates into the atmosphere. Because the water in the closed basin was mostly isolated from other streams and rivers, developers got the idea that they could appropriate "non-beneficial evapotranspiration" by "salvaging" water. From there, they would deliver the salvaged water to irrigators for use in meeting Colorado's obligations under the Rio Grande Compact. Thus was born the Bureau of Reclamation's Closed Basin Project that Congress authorized in 1972.[86] To get at this water, the Bureau of Reclamation drilled multiple groundwater wells as part of what they called the "Closed Basin Drain" that convey the produced water about 42 miles through a canal to the Rio Grande River. By drilling these wells, the Bureau of Reclamation "salvaged" the "non-beneficial evaporation" consumed by wetland plants in the closed basin. By lowering the groundwater levels below the root depth of wetland plants, the Bureau of Reclamation succeeded in drying up thousands of acres of wetlands. In any given year, the project delivers about 43,000 acre-feet of water to the Rio

Grande Water Conservation District. Critics of the Closed Basin Project decry the dried-up wetlands and the lowered aquifer levels for the impact on local ecosystems and riparian habitat, streamflow, and water quality.[87]

CLOUD SEEDING

For water-starved farms and industries, the prospect of creating more rainfall or increasing snowpack by any means possible is alluring. Scientists have long experimented with weather modification by seeding clouds in an attempt to enhance precipitation. Most cloud seeding work has centered on increasing winter snowpack. Since at least the 1880s, people have attempted to modify the weather and generate rain by shooting "hail cannons," firing rockets, and ringing bells.[88] Then, in the 1940s, researchers discovered that dispersed silver iodide could prompt the nucleation of ice crystals in supercooled water vapor. The silver iodide seems to work because its structure appears to resemble that of ice crystals.[89] This pioneering laboratory research suggested that if engineers inject dust-sized particles of silver iodide or similar substances into the atmosphere, they might induce the consolidation of microscopically dispersed water molecules. As water congeals around the dust, droplets grow and eventually fall to the earth as snow or rain. The practice of cloud seeding in summer versus winter

Figure 2.47. Here a plane flying over North Dakota fires a flare to seed clouds. This is part of the weather modification program managed by the North Dakota Atmospheric Resource Board. Image courtesy of the North Dakota Atmospheric Resource Board. Source: https://www.swc.nd.gov/info_edu/galleries/Weather%20Modification%20(ARB)/index.php.

is very different. In the summer, cloud seeders use salt to condense water and grow large droplets. In the winter, silver iodide works better for producing snow. Cloud seeders disperse silver chloride or other salts from the ground or airplanes into humid air to stimulate droplet formation.[90]

Scientists have long debated the effectiveness of cloud seeding. Still, water managers involved with cloud seeding programs maintain that the rain or snow it produces can alleviate drought or augment water supplies. In recent years, scientific consensus has shifted from questions of whether cloud seeding even works to how well it works.[91] Recent research suggests that cloud seeding may indeed increase precipitation, albeit by minute amounts.[92] Even so, water providers have found the work intriguing enough that they regularly include cloud seeding projects within programs to augment water supply. In 2018, for example, Colorado River Basin states eager to enhance snowpack agreed to shift work from individually funded projects to a multistate collaborative program.[93] It seems clear that work on cloud seeding and weather modification will continue in earnest. With the adverse effects of climate change becoming more pronounced and accelerating, water managers will likely support cloud seeding projects to hedge against the risk of further aridification in the Southwest.

COFFIN V. LEFT HAND DITCH COMPANY

Since the beginning of the Colorado Gold Rush in 1858, Euro-Americans grappled with how to efficiently allocate water in the region's dry climate. They quickly concluded that the legal doctrine of riparian rights in use in the humid eastern states was wholly inadequate in this semiarid land. So, colonists needed to create a system better suited to the western climate. In short order, they enacted laws that would allow immigrants to claim water from rivers and divert it to their land, even if it meant digging ditches across private or public property. Colonists also recognized that the wide distribution of water was necessary for development, and conversely that consolidating water rights into the hands of corporations or a few wealthy individuals would stymie the territory's growth. As a result, when Colorado became a state in 1876, colonists adopted constitutional provisions to encourage water development. One clause declared that all water was the property of the state and another that the "right to divert shall never be denied." Even with a new constitution, there remained uncertainty about how to implement these provisions. What they needed was a test case.[94]

The opportunity came in the exceedingly dry summer of 1879. Many farmers found their crops wilting. Among them was Rubin Coffin, a Civil War veteran

Figure 2.48. The Left Hand Ditch diverts virtually the entire flow of Left Hand Creek through a system of laterals that are spread out along the creek. Here is a headgate for one of these laterals in 2011. Author photo.

who lived along St. Vrain Creek near present-day Longmont, Colorado. Coffin resented that the Left Hand Ditch Company had a diversion dam in the high country near the Continental Divide and that the company was diverting water from South St. Vrain Creek into a tributary of Left Hand Creek. Coffin gathered some neighbors and rode up to the mountains and tore out the Left Hand Ditch Company diversion. A bitter feud ensued. The ditch company staged armed guards to protect the diversion, and this feud started the lawsuit that would set the terms for western water law for the next century.

A trial took place in Boulder, Colorado, and Coffin and his neighbors were quickly found guilty of destroying the diversion dam. They appealed the verdict to the Colorado Supreme Court. In 1882, the court upheld the ruling of the lower court and then affirmed the principles of the Prior Appropriation Doctrine. In the Colorado Supreme Court decision, the justices once and for all extinguished the English common law practice of riparian rights. They also upheld the "first in time, first in right" system of water appropriation. They recognized the legal right of a water diverter to take water from one watershed to another. Importantly, the Colorado Supreme Court also allowed water users to irrigate lands away from a stream and permitted a water user to establish a right of way across private, corporate, or public lands to access the water. This court decision ended up becoming one of the foundational decisions that let water rights law develop in a manner consistent with the arid conditions prevalent in the western United States. Taken as a whole, this decision, and laws associated with it, became what we now call the Colorado System of Prior Appropriation, or more generally in the West, the Prior Appropriation Doctrine. Eventually, every western state adopted in whole or in part the principles articulated by the court in *Coffin v. Left Hand Ditch Company.*

COLORADO RIVER

For at least 12,000 years, Indigenous people lived and died along the banks of the Colorado River and its tributaries. Major precontact civilizations grew and declined within its basin. The great Chacoan culture and their descendants, the Fremont peoples, the Hohokam, among others, thrived along the Colorado River and its tributaries. Modern tribes, including the Pueblo peoples, the Navajo, the Tohono O'odham, and many others, still call the Colorado River Basin their home.

Europeans first set eyes on the Colorado River when Francisco de Ulloa sailed to the head of the Gulf of California in 1536. During Coronado's 1540–1542 failed expedition seeking the "Seven Cities of Gold," he sent a party led by García López de Cárdenas to modern-day Arizona where they became the first Europeans to see the Grand Canyon. The river got its modern name, the *Rio Colorado* ("Red River"), when Father Eusebio Francisco Kino produced maps and reports from his overland explorations in 1700–1702 to the Colorado River Delta.

Americans began exploring the Colorado River and its tributaries during the early nineteenth century. Jedediah Strong Smith traveled to the lower Colorado River canyon in 1826, seeking beaver pelts. Many others soon followed. Chief among them was Joseph C. Ives, who commanded an expedition in 1857 to 1858 up the Colorado River from its mouth. Ives designed and built a sternwheel steamboat and shipped it to the Colorado River Delta, where his party then ascended the river. Using his 54-foot steamboat *Explorer,* the expedition ascended the lower Colorado River for 550 miles into the Black Canyon of the Colorado River, an area now submerged by Lake Mead. John Wesley Powell completed the initial explorations of the river during his famous 1869 descent of the river in small boats.

With the advent of the Reclamation Age, the Colorado River became a prime target for water developers. Today the main stem of the river has fourteen major

Figure 2.49. This is a boater flipping his raft in Lava Falls, the largest and most famous of all rapids on the Colorado River, and for that matter of any other river in the American West. Photo courtesy of the author.

dams, and tributary streams have another thirty-one. Collectively, these projects make the Colorado River one of the most regulated watercourses in the United States.[95]

In 1921, at the urging of Colorado Congressman Edward T. Taylor, the US House of Representatives and Senate approved a resolution changing the name of the Grand River to the Colorado River. Taylor argued that the Grand River, a tributary of the Colorado River that flows through the states of Colorado and Utah, should get renamed as the Colorado River to match the state containing its headwaters. Utah and Wyoming politicians objected to the change, arguing that the Green River, which crosses those states, is the longer tributary and, therefore, should be afforded that honor. Taylor countered that there are other precedents in which a shorter tributary bears the name of the master river (for example, the Mississippi River has its name on a shorter branch, whereas the longer tributary, the Missouri River, retains its name). Taylor's arguments prevailed, and Congress officially changed the name of the river to match the state of its origin. Consequently, contemporary maps show the Colorado River extending from headwaters in Colorado's Rocky Mountain Park to its delta in the Gulf of California. This name change also explains why there is a Grand County and a City of Grand Junction in Colorado.

Figure 2.50. To promote the Colorado River Aqueduct, a group called the Citizens Colorado River Water Committee produced about 500,000 copies of this promotional brochure in 1931 and had them inserted in water bills in Southern California. Image courtesy of the Metropolitan Water District of Southern California. Public domain image.

COLORADO RIVER AQUEDUCT

One of the most significant water delivery structures of the American West, the Colorado River Aqueduct, begins at the Parker Dam along the California/Arizona border southeast of Lake Havasu City, Arizona, and delivers water to its terminus at Lake Mathews in western Riverside County, California. Water flowing within the aqueduct passes through 90 miles of tunnels, 5 pumping stations, approximately 55 miles of cut-and-cover conduit, and almost 30 miles of siphons.

The project was so vast and ambitious that the American Society of Civil Engineers named the aqueduct one of the world's Seven Modern Civil Engineering Wonders. William Mulholland of Los Angeles Water and Power initially conceived the project. But the project was designed and built under the supervision of Frank E. Weymouth and the Metropolitan Water District of Southern California (MWD). Construction took place between 1933 and 1941; employing approximately thirty thousand people, the Colorado River Aqueduct was the most extensive public works project in Southern California during the Great Depression.

With the capacity to deliver approximately 1,600 cfs of water, the aqueduct enabled much of the startling urban sprawl that consumed Southern California. In the years following its completion, the six counties served by the MWD grew in population from 3.5 million in 1939 to 17.5 million in 1993.

COLORADO RIVER DELTA

Although known to the Indigenous people for millennia, Europeans first encountered the Colorado River Delta during Hernando de Alarcón's explorations in 1540. In the aftermath of the Spanish conquest of Mexico, the delta

Figure 2.51. Shown here is a satellite view of the Colorado River Delta acquired from the Operational Land Imager (OLI) on Landsat 8 from March 20, 2020. Surrounding the delta to the west are white salt flats, and the shifting brown dune fields of the Sonoran Desert extend to the east. In the center of the photograph is Montague Island. Public domain image courtesy of NASA. Source: https://earthobservatory.nasa.gov/images/146839/green-lagoons-no-more.

remained a sparsely populated and seldom visited area. But change was on the horizon. Through a quirk of nineteenth-century diplomacy in the aftermath of the Mexican-American War, Mexico ceded the vast majority of the Colorado River Basin to the United States. However, the last few miles of the river, including its delta, remained part of Mexico. By the time Aldo Leopold described his 1922 canoe trip to the delta in his acclaimed book, *A Sand Country Almanac*, twenty-seven years after the fact, the delta was in steep decline. Even then, what Leopold saw in 1922 scarcely differed from the landscape Alarcón encountered centuries before. Leopold described a "verdant wall of mesquite and willow [that] separated the channel from the thorny desert beyond. At each bend we saw egrets standing in the pools ahead, each white statue matched by its white reflection. Fleets of cormorants drove their black prows in quest of skittering mullets; avocets, willets, and yellowlegs dozed one-legged on the bars; mallards, widgeons, and teal sprang skyward in alarm. As the birds took the air, they accumulated in a small cloud ahead, there to settle, or to break back to our rear. When a troop of egrets settled on a far green willow, they looked like a premature snowstorm."[96]

This Arcadian scene, as spectacular as it was, was perhaps second to the delta's other remarkable feature: a tidal bore of extraordinary strength. Created by the counterforces at play between the river flow and the ocean tide, a powerful wave occasionally formed and migrated upstream. Joseph C. Ives, surveying for the US Corps of Topographical Engineers, witnessed the bore on the evening of November 30, 1857:

> While the tide was still running out rapidly, we heard . . . a deep, booming sound, like the noise of a distant waterfall. Every moment it became louder and nearer, and in half an hour a great wave, several feet in height, could be distinctly seen flashing and sparkling in the moonlight, extending from one bank to the other, and advancing swiftly upon us. While it was only a few hundred yards distant, the ebb tide continued to flow by at the rate of three miles an hour. A point of land and an exposed bar close under our lee broke the wave into several long swells, and as these met the ebb the broad sheet around us boiled up and foamed like the surface of a caldron, and then, with scarcely a moment of slack water, the whole went whirling by in the opposite direction. In a few moments the low rollers had passed the island and united again in a single bank of water, which swept up the narrowing channel with the thunder of a cataract. At a turn not far distant it disappeared from view, but for a long time, in the stillness of the night,

the roaring of the huge mass could be heard reverberating among the windings of the river.[97]

With the construction of the great Colorado River Dams, starting with the Boulder Canyon Project, and continuing with the multitude of diversions siphoning water away from the Colorado, the delta itself has all but disappeared in the absence of flow, and the bore has become a distant memory.[98]

COLUMBIA RIVER

As the largest river in the western United States by volume (outside of Alaska), the Columbia holds an outsized place in America's history and imagination. Clocking in at over 1200 miles long, the Columbia River rises in the Rocky Mountains of British Columbia and flows south into Washington state, and eventually forms the border of Oregon and Washington. For at least fifteen thousand years, Indigenous people lived and died along the Columbia's banks, fishing for salmon and hunting the abundant wildlife that the river supports. The Spaniard Bruno de Heceta (Hezeta) y Dudagoitia became the first known European to see the Columbia when his expedition sighted the river's mouth in 1775. Occasional Spanish explorers passed by, but none entered the river. Then in April 1792, American captain Robert Gray saw the mouth of the Columbia River, and even spent nine days trying, but ultimately failing, to cross a treacherous bar that separated the river and ocean. He returned later that May and succeeded in crossing the bar, becoming the first person of European descent known to enter the river itself. That same spring Royal Navy Captain George Vancouver sailed past the Columbia's mouth but did not enter it. His sighting provoked a decade's long rivalry between the United States and Great Britain (at times entangling Russia and Spain) for control of the Columbia and the fur trade that the river supported. Lewis and Clark famously traveled this river in their journey across the American continent in their expedition to seek the fabled Northwest Passage at the behest of Thomas Jefferson in 1804. Competition for control of the Columbia continued until the United States and Great Britain concluded the Oregon Treaty of 1846 that set the boundary between Canada and the United States at the forty-ninth parallel.

More recently, the historian Richard White aptly described the Columbia as an "organic machine" and asserted that the river's human history and natural history are so intertwined that understanding either is impossible without the other.[99] Indeed, people have harnessed the Columbia River as a hydrological machine to

Figure 2.52. This photograph, taken from the Columbia River Highway in 1908, shows the steamboats *Bailey Gatzert* (left) and the *Charles R. Spencer* in rapids approaching the Cascade Canal and Locks. The image hints at the massive volume of water that flows in the Columbia River. Photographer unknown. Image courtesy of the Oregon Historical Society. Source: https://www.oregonhistoryproject.org/articles/historical-records/steamboats-on-the-columbia-river-at-cascade-locks/#.YJR2_y1h2X2.

power a vast network of hydroelectric dams that sustains the northwest United States economy. The Columbia's ecology, once home to seemingly unlimited numbers of salmon, was sacrificed to underwrite the water development. Today, complex politics pit native peoples, the hydroelectric industry, a broader public sympathetic to the river, a tech industry hungry for power, and environmental groups wishing to restore salmon against one another for the Columbia River's water.

COLUMBIA RIVER MEGAFLOOD AND GLACIAL LAKE MISSOULA

A virtually unimaginable amount of water cascaded across central Washington when glacial lake Missoula breached during the Pleistocene Ice Ages. Estimates of the flood near its point of release in the Spokane Valley suggest that the peak discharge probably exceeded 600,000,000 ± 105,900,000 cfs. At the Wallula Gap, a narrow reach of the Columbia River downstream from Spokane where floodwaters converged, flows may have exceeded 353,000,000 ± 88,000.000 cfs. To give

Figure 2.53. This satellite photo shows the distinctive channeled scabland terrain of Eastern Washington. With north at the top, the northeast-southwest trend of the hills is a result of erosion and deposition from the Columbia megaflood. The town of Dayton, Washington, is in the southeast part of the image. This flood was so vast that it takes satellite images such as this one to see the distinctive landforms created by the flood. Image source: screen capture Google Earth.

a feeble visualization of what this entailed, the highest discharge recorded in historic times on the Columbia River at The Dalles was approximately 1,160,000 cfs or a mere 2/1000ths of the glacial lake flood flow. This flood was so gigantic that for decades after geologist J. Harlen Bretz first published papers describing evidence for the flood, fellow geologists ridiculed Bretz claiming that such a cataclysm was impossible. It took the advent of the space age and the acquisition of satellite images in the 1970s to shift the geological consensus towards Bretz's view. Even then, it was many more years before sufficient geological evidence and theoretical analysis produced a clearer picture of this Pleistocene calamity. Fortunately, Bretz lived long enough to see his views vindicated, and in 1979, when he was 96, he was awarded the Penrose Medal, the Geological Society of America's highest award.

Geologists now believe that the lake breach was a form of jökulhlaup, the Icelandic term for a type of glacial outburst flood. These breaches occur when a glacial ice and moraine dam back water up into a lake behind it. Eventually, the ice and moraine dam destabilizes and fails, releasing vast amounts of water downstream. For the Channeled Scabland of Washington, the ice dam's rupture was probably so complete that a wall of water as much as 1,200 feet high issued out from the lake to reconfigure the Columbia's watershed. The flood itself happened about 12,000 years ago, so it is likely that some of the Indigenous people living in the region witnessed the event and were perhaps even swept away by it. In satellite images of the scablands, one can quickly identify features of the flood, including northeast-southwest trending lakes, alluvial deposits, and the flood carved "scabland" terrain.[100]

CONFLUENCE OF THE GREEN AND COLORADO RIVERS

After passing the junction of the Green and Colorado (then known as the Grand) Rivers, the 1869 Powell expedition stopped and camped for a few days to regroup. Major Powell and expedition member George Y. Bradley set out to climb up the desert cliffs to view the junction: "And what a world of grandeur is spread before us!" Powell exclaimed. "Below is the canyon through which the Colorado runs. We can trace its course for miles, and at points catch glimpses of the river. From the northwest comes the Green in a narrow winding gorge. From the northeast comes the Grand, through a canyon that seems bottomless

Figure 2.54. On this satellite photo, you can clearly see the confluence of the Green (upper left) at its junction with the Colorado River in Canyonlands National Park. Notice how the Green River dominates the combined river by assimilating the red silt of the Colorado below the confluence. Screen capture of image courtesy of Google Earth.

from where we stand. Away to the west are lines of cliffs and ledges of rock—not such ledges as the reader may have seen where the quarryman splits his blocks, but ledges from which the gods might quarry mountains that, rolled out on the plain below, would stand a lofty range; and not such cliffs as the reader may have seen where the swallow builds its nest, but cliffs where the soaring eagle is lost to view ere he reaches the summit."[101]

The junction of the Colorado and Green Rivers, within the rugged desert country at the center of Canyonlands National Park, is the very heart of the American West. It is spectacular but desolate, difficult to reach by all but the most intrepid hikers or boaters. Through the accolades of Powell and others who followed, this river junction symbolizes the soul of America's Southwest. Although there are countless other river junctions across the region, none better represents the grandeur of western rivers, wildness, and raw natural beauty.

CONSERVATION AND CONSERVANCY DISTRICTS

Apart from the occasional highway sign announcing that one is crossing the boundary into or out of a conservation or conservancy district, most Americans remain unaware that these quasigovernmental organizations even exist. Yet, across the West, a multitude of conservation and conservancy districts work on a diverse range of local soil and water issues. Generally, conservation and conservancy districts have very similar missions and policies. The main difference is that concerned landowners form conservancy districts and conservation districts get formed through legislative enactment.[102] In Wyoming alone, for example, thirty-four conservation districts work on issues important to locals. Many districts were formed in the aftermath of the Dust Bowl as ways to recover from the drought by stabilizing rangeland and protecting and developing water resources. For the founders of these quasigovernmental entities, conservation of water often meant storing and diverting water so that local interests might use it. In those days, allowing a stream to flow in its natural state was considered a waste of water, hence the need to "conserve" it. This perspective led the journalist Ed Quillen to quip that a "conservancy district is a legal device for destroying fisheries, riparian habitats, wetlands and indigenous populations."[103] Today, some districts take a broader view of water management to include environmental uses and water quality protection. As quasigovernmental entities, conservation and conservancy districts elect board members from within their service areas and use their taxing authority (usually via property taxes) to support

Figure 2.55. Signs advertising conservation and conservancy districts dot the Western landscape. Here is one for the Lower Wind River Conservation District in Wyoming. There are 34 conservation districts in Wyoming alone. Photo courtesy of the author.

their programs. Over the years, activities of these districts have expanded to include education, environmental applications such as recreational stream flow management, water quality management, tree and shrub sales, rangeland management, and soil testing. A central role for these districts is to represent local interests with state and federal agencies.

CONSUMPTIVE USE

In the nineteenth century, Americans led the world in seeking ways to monetize nature. That century brought us grain elevators, pork belly futures, and commodity markets. In the American West, colonists vigorously extracted gold and

Figure 2.56. The consumptive use of water on irrigated fields varies depending on the type of crop and soils that are present. This alfalfa field near Loa, UT, is being spray irrigated. Irrigation by this method is more efficient than flood irrigation but less efficient than drip irrigation. Photo courtesy of the author.

other minerals, cut forests, killed bison, and appropriated all the water that they could. Gold and buffalo were relatively easy to take, but water, a fluid that could seep into the ground or evaporate, posed particular problems. To fully monetize water and make it a commodity—something that they could buy and sell—users had to determine how much water they used. For this reason, they made great strides in determining flows and volumes. To fully understand how much water they used, they also needed to calculate how much water crops consumed and how much soaked into the ground to flow back to the river.

These practical needs for determining how much water a plant uses led engineers to the notion of consumptive use. Said differently, water is evaporated from ditches, transpired by plants, incorporated into crops or other products, used as drinking water by humans or livestock, or somehow deliberately removed from the stream system, all forms of consumptive use. So, for example, a farmer will divert water from a stream into her headgate, run it down a ditch to the farm, irrigate her crops, and then allow the leftover water to flow back to the creek. As water moves down a ditch, some evaporates from the channel, more seeps into the ground, additional water evaporates from the sprinkler system, and some is taken up by the crop and incorporated into the plant cells. It is this water that engineers refer to as consumptive use. The remaining water flows back to the stream, where other users wait for it. Consumptive use, therefore, is always less than the amount diverted. And it is the consumptively used portion of a water right that is bought and sold.

Consequently, attorneys and engineers pay careful attention to all the measurements and mathematics that quantify consumptive use. It is consumptive use that often determines the amount of water in a water right decree. Downstream users likewise carefully monitor the consumptive use of others. If, for example, an upstream user installs water conservation facilities that allow him to grow more crops and wind up consuming more water relative to the amount he historically diverted, the downstream user may object to the reduced flows returning to the creek. Water managers call these new diversions an "expansion of use," meaning that in a fully utilized stream somebody (or something such as fish) somewhere ends up getting less water than they were historically entitled to receive.

COOPERATION

In a world where conflict makes headlines and dire predictions dominate the op-ed pages, it is no surprise that many people perceive western water as fraught with stubborn problems and intransigence between opposing parties. It doesn't help that many economists have made gloomy predictions about resource consumption outstripping supply in everything from food to water. Perhaps no economist since Thomas Malthus has reinforced the notion that economics is the "dismal science" better than Garrett Hardin and his *Tragedy of the Commons*. Hardin essentially asserts that a libertarian free-for-all by rational, utility-maximizing individuals over common-pool resources will inevitably lead to depletion and collapse.[104] It is tempting to cite Hardin's essay when considering groundwater depletion across the West or the overappropriated Colorado River. Yet, for the most part, Hardin's somber predictions have not come to pass.

In contrast to Hardin, look to the work of Elinor Ostrom. She demonstrated that the tragedy of the commons is only one endpoint in a spectrum of possible outcomes ranging from noncooperation to cooperation. Ostrom challenged Hardin's approach to the *Tragedy of the Commons* by advocating bottom-up management by individuals and communities to allocate collective resources. As Ostrom put it, "Pessimism about the possibility of users voluntarily cooperating to prevent overuse has led to widespread central control of common-pool

Figure 2.57. The state and federal representatives shown here negotiated the Colorado River Compact at Bishop's Lodge, NM, in 1922. Although conflict sells newspapers, in the long run, people have solved many of the most difficult water issues in the West through cooperation and compromise. Public domain photograph courtesy of the US Bureau of Reclamation.

resources. But such control has itself frequently resulted in resource overuse. In practice, especially where they can communicate, users often develop rules that limit resource use and conserve resources." Ostrom, the first woman to win a Nobel Prize in Economics, for, among other things, her work on cooperative efforts to address groundwater depletion in California, convincingly argues that cooperation and collaboration ultimately provide the greatest chance for resolving thorny depletion issues.[105]

Indeed, it is Ostrom's perspective that best aligns with the history of water development in the West. Since people began building water projects, cooperation, collaboration, and communication have remained the essential recipe for success. In what is now Arizona, Hohokam people could not have constructed their vast waterworks without careful organization, planning, management, and labor mobilization. Likewise, the Hispanic peoples of the Southwest erected their waterworks and distributed water through the close cooperation of entire communities. Moreover, the Hispanic system of water sharing, or *reparto*, during times of drought exemplifies the success of collaborative and coordinated management over many generations. And if one looks past the debunked myth of the rugged Western individuals, we see that Americans also built their waterworks and water institutions through close cooperation, coordination, and communication. Consider mining district bylaws, ditch building, and water administration, and it is clear that the overriding theme is negotiation, coordination, and compromise over solo effort or conflict.

That is not to say that we have not had conflict and tension. We have. But we have had far more issues resolved through negotiation and compromise. And, I maintain that my assertion is borne out by the evidence. For example, during my time managing water resources in Boulder, Colorado, I observed that fewer than 5 percent of cases filed in Water Court ever ended in a trial. The vast majority of cases concluded in negotiating sessions and not in front of a judge. Moreover, far more water issues were settled by talking things over and finding solutions as opposed to filing a suit. Likewise, the legacy of negotiating interstate river compacts and their implementation shows us that if there is a will there is a way even for what appear to be the biggest and most stubborn problems.

COULEE

A French-Canadian word that has jumped to English, *coulee* refers to small streams, dry streambeds, or small-shallow ravines. The word's etymology is a derivation of the French word *couler* meaning "to flow," and before that, from the Latin *colare*, meaning "to strain or flow." In the American West, it is sometimes used synonymously with the word "gully." Because of its French origins, we see it used in the western United States in places where French or French-Canadian trappers once roamed. In particular, the word coulee is commonly encountered on maps in the Northwest and in the Mississippi Delta regions.

Figure 2.58. View of Grand Coulee, below Dry Falls, WA. The blocky-looking cliff-forming layer on the upper part of the canyon is basalt lava flows. Dry Falls itself formed during the massive Columbia megaflood of the Pleistocene. (Photograph source: Wikipedia, public domain.)

CUBIC FEET PER SECOND

Whenever you want to buy or sell a commodity, it's critical to have an accurate method for calculating how much of the product you have. Water is no different, and calculating water quantities is particularly critical in the American West. Measurement is relatively easy for products like steel or grain, but liquid water is elusive; it flows away, seeps into the ground, and evaporates.

Because accurate measurements for water quantity are needed, an otherwise obscure number, *cubic feet per second*, became widely adopted in the nineteenth century to record the flow rates of rivers, streams, and ditches across the region. Cubic feet per second (writers sometimes shorten it to *second feet* or abbreviate it as *cfs*) is a rate of flow and is equal to a volume of water 1 foot high and 1 foot wide moving a distance of 1 foot in 1 second. One cubic foot contains 7.48 gallons of water. As an example, if you have a bathtub that is 2 feet wide by 1 foot across and

Figure 2.59. During the nineteenth and early twentieth centuries, one of the significant problems facing water managers was how to measure the flow of water accurately. Although many engineers considered the question, it fell to Ralph Parshall, shown here in front of a flume of his design, to make the most significant lasting contribution toward resolving this issue. His solution was to design a standardized flume and recording devices that made it easy to calculate the flow of water, typically given in cubic feet per second. Water users employ flumes of his design across the West to this day. Photo courtesy of Colorado State University.

1 foot deep (2 cubic feet), then water from a spigot flowing into it at a rate of 1 cfs would fill the tub in two seconds. By standardizing water measurements around cubic feet per second, it became possible to determine how much water a person might be claiming and also provide a basis for eventually selling that water.

Perhaps you are not the only one who has noticed how wonky and awkward the value cfs is and all the odd mathematical conversions associated with it. You can thank the unwieldy and outdated British Imperial measurement system used here in the United States for that. In a roundabout way, we base our streamflow measurements on an obscure British monarch's foot size. America was once on track to jettison the system and replace it with the infinitely more straightforward and rational metric system (you only need to know how to divide by ten to use it), but unfortunately, nationalistic politics intervened. Demonizing the metric system as a rouse by foreigners (it didn't help that pesky French socialists devised the metric system) to undermine American competitiveness, reactionary elements in the US fought its adoption. Enlightened scientific and economic powerhouses, such as the National Cowboy Hall of Fame, sued to block the adoption of the metric system, claiming "the West was won by the inch, foot, yard, and mile." President Ronald Reagan, bowing to his xenophobic supporters, scuttled the conversion effort in a federal budget bill. So, here we remain straddled with the outdated system in government reports and gages, even though converting to metric would simplify everyone's mathematics, and be cost-effective to boot.[106]

DEBRIS FLOWS

John McPhee, America's best-known writer about geology, recounted the terror of being engulfed by a debris flow during an event near Los Angeles in the 1980s:

> Large tree trunks rode in the debris like javelins and broke through the sides of houses. Automobiles went in through picture windows. A debris flow hit the gym at Azusa Pacific College and knocked a large hole in the upslope wall. In the words of Cliff Hamlow, the basketball coach, "If we'd had students in there, it would have killed them. Someone said it sounded like the roar of a jet engine. It filled the gym up with mud, and with boulders two and three feet in diameter. It went out through the south doors and spread all over the football field and track. Chain-link fencing was sheared off—like it had been cut with a welder. The place looked like a war zone."[107]

Although sometimes stunning in their destructiveness, debris flows are a common occurrence in the mountains and canyons of the West. They are just

Figure 2.60. Debris flows are a significant vector of landscape change in the American West. They often occur during major storms that have saturated surrounding hillslopes. The risk for debris flows becomes particularly acute after wildfire events reduce the vegetation cover that helps stabilize hillsides. For example, debris flows generated in Mullally Canyon damaged this house in response to a rainstorm on February 6, 2010. During the previous summer, the drainage basin above this home was burned by the Station Fire, the largest fire in the history of Los Angeles County. Photography by Susan Cannon, courtesy of the US Geological Survey. Public domain image.

one of a wide variety of hillslope failures that result in sediment's mass movement. When these dramatic events strike, they can cause widespread casualties and catastrophic property damage. Scientists categorize hillslope failures into debris avalanches, earthflows, mudflows, solifluction,[108] and lahars.[109] All of these hillslope failures tend to be variations of each other due to interrelated factors such as sediment type, sediment size distribution, moisture content, hill slope angle, rate of failure, and so on. Debris flows tend to possess the character of viscous fluids. Entrapped air, water, and clay facilitate flow and help create the buoyancy that floats giant boulders and fully grown trees down the hillside. To give you a sense of the vast power of a debris flow, consider the 1928 St. Francis dam collapse. When the dam failed, the ensuing flood created a human-made equivalent of a debris flow that lifted an enormous concrete fragment of the dam—nearly 60 feet across and weighing 10,000 tons—and moved it 1,500 feet downstream from its original position. In areas susceptible to hillslope failure, events such as wildfire that incinerate trees and undergrowth can reduce local slope stability. That coupled with rain, in particular monsoonal-type downpours that saturate the soil, further reduces stability until a threshold is reached and the whole mass of the hill begins moving. Poor zoning and the desire of people to live in scenic, but vulnerable, areas exacerbate the danger of debris flows.

DELTA SMELT

If there has ever been a fish that has had proper legal representation, it is the delta smelt (*Hypomesus transpacificus*). This silvery translucent fish lives in California's Sacramento–San Joaquin Delta—the very center of California's water distribution system—making this tiny fish a key player in the region's hydropolitics by virtue of the environmental community's keen interest in its fate. Biologists fear that the delta smelt faces extinction in the wild. Its most significant threat is the lack of freshwater flows to sustain its habitat. But it doesn't help that the water it lives in is polluted by pesticides and other toxic chemicals. Plus, as the populations of the fish continue to decline, its genetic diversity is becoming undermined, making prospects for recovery particularly dicey.

As much as 50 percent of all freshwater that formerly reached the delta now gets diverted for agricultural, industrial, and municipal uses. Compounding this is the severe drought, and declining snowpack has led to reduced flows in the Sacramento River over the past several years. All this means that the delta smelt, which depends on slightly saline areas of the delta for survival, is struggling with unhealthy levels of salinity across essential parts of its range. As of 2015, fisheries biologists found only six delta smelt during an annual netting survey. Nevertheless, biologists estimate that about 48,000 survive in the wild, which is only a tiny fraction of numbers that once inhabited the delta. Because the species is so imperiled, in 1993 it was listed as threatened under both the federal Endangered Species Act (ESA) and the California Endangered Species Act (CESA). In 2009, as the species numbers continued to decline, the CESA status was changed to endangered.[110]

Each adult delta smelt is tiny: they typically grow to about 60–70 mm (or a little longer than the short side of a credit card) and occasionally reach 120 mm when fully mature. Their life cycle follows the seasons. They spawn in spring in the freshwater of the upper river and then migrate downstream during the summer to the less salty areas of the delta. Here they remain until the fall, where they

Figure 2.61. This is the delta smelt, a small fish with large political clout. Public domain image courtesy of the California Department of Water Resources. Photo by Dale Kolke. Public domain image.

mature. They then complete their seasonal cycle in the winter when they head back upstream shortly before spawning. Reversing the delta smelt's march toward extinction will require major shifts in California water policy and water use to ensure that enough freshwater is present in the delta to manage salinity levels in critical habitat. It will also require ongoing active management of the smelt's habitat and life cycle and a sustained political commitment to avoid its demise.

DENDROCHRONOLOGY

One of the enduring questions in archaeology is, Just how old is that? Until the early twentieth century, the best that archaeologists could do was to establish the relative dates that cities were built or abandoned. Information on exact timing remained frustratingly elusive, particularly in places where no written record existed to confirm a date.

Dendrochronology started as a research project to provide precise dates when archaeological sites in the Southwest were constructed and deserted. The term is a combination of the Greek words for tree, time, and study. By counting and analyzing the thickness and structure of annual tree rings and the relative

Figure 2.62. The astronomer A. E. Douglass explains tree-ring dating to a group of visitors at Arizona State Museum. Charles Herbert, photographer, c. 1940–1950. Image courtesy of the Arizona State Museum, University of Arizona. Image no. 18493.

patterns of thin and thick rings, it is possible to determine the exact year that a particular tree ring grew. Information from this discipline has allowed archaeologists to determine absolute dates for cutting trees used in ancient structures. Because there are close relationships between tree rings' relative thickness to annual moisture, scientists can infer wet and dry climatic cycles. This correlation led to the emergence of the closely related fields of dendrohydrology and

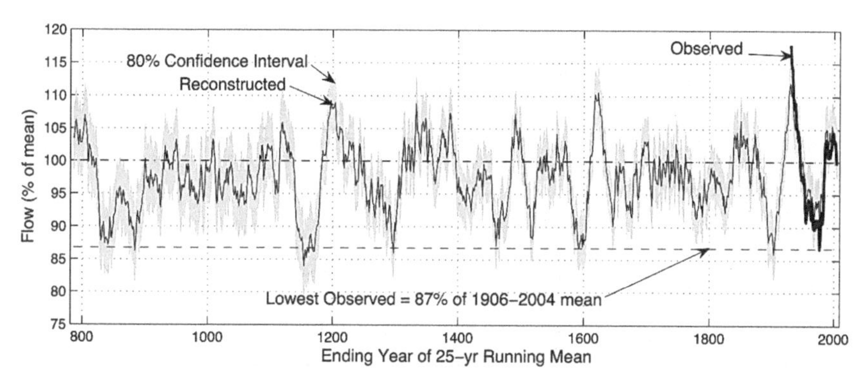

Figure 2.63. This is an updated version of the groundbreaking "Stockton chart" showing Colorado River flow from the years 762 to 2005. The chart, a hydrological reconstruction of streamflow based on tree ring widths calibrated to gage data from the river, indicates that some of the wettest years in the past millennium were the very years used to apportion water for the Colorado River Compact. The chart shown here is courtesy of David M. Meko.

dendrogeomorphology. Respectively, these disciplines use tree ring data to reconstruct climate and hydrology patterns and better understand landforms' evolution.

An astronomer, Andrew Ellicott Douglass, single-handedly pioneered the field. Douglass (1867–1962) coined the term while working at the Lowell Observatory near Flagstaff, Arizona. Douglass initially collected tree samples to determine if he could see a link between climate and sunspot activity. Douglass quickly realized its potential for dating archaeological sites, a revelation that revolutionized archaeology. In due time, researchers developed correlations between tree ring width and streamflow that allowed them to estimate flows for rivers in the years before engineers erected stream gages. This research led to startling discoveries of deep droughts and climate variability in the centuries before the American occupation of the Southwest.

The field of dendrochronology led to a remarkable advance in understanding southwestern hydrology with the publication of what is commonly known as the "Stockton chart," named after its author.[111] It resulted from a groundbreaking correlation between tree rings and discharge on the Colorado River. The primary result, which has not substantially changed since its publication, is that the approximately twenty years preceding the Colorado River Compact's ratification in 1922 were some of the wettest on record. This result was a significant "uh oh" moment for western water managers. They realized that the Colorado River Compact's hydraulic assumptions were much more optimistic than the historic record indicated, a situation corroborated in the decades since the ratification of the compact. More seriously for the Southwest, tree ring data also supports research that human-induced climate change will lead to significantly worse droughts than the region experienced over the past several thousand years. Sadly, when scientists link the dendrochronology data to rigorous climate models, researchers come to the troubling conclusion that the water future for the Southwest looks bleak.[112]

DESALINIZATION

People have sought ways to remove salt from water for generations. If we can do it economically, we might vastly expand our water supplies. The process, called *desalinization*, is the removal of salts from water to provide fresh water. Heralded as a potential solution for the West's water deficits, there is a wide gap between the desire of what desalinization might achieve and the reality of what is happening. Water managers generally consider freshwater as containing up to about 1000 parts per million (ppm) of dissolved salts. When the salinity of water exceeds that level, it is no longer fit for human consumption and can also harm plants. Seawater is much saltier and contains about 35,000 ppm of dissolved salt. Many technologies exist for removing salt from water, and some are quite ancient. Perhaps the oldest known desalinization technology is distillation

Figure 2.64. Desalinization membrane installation at the Alameda County's Newark Water Desalination Facility. Public domain image courtesy of the California Department of Water Resources. Photo by John Chacon.

whereby water is heated (often to boiling point) to evaporate water from one container containing salty water and condense into another without the salt. A solar distiller works similarly but is heated by sunlight to make the water evaporate, providing salt-free condensate for use.

Another widely employed method is through osmosis, in which water diffuses through a semipermeable membrane leaving the salt behind. Commercial-scale desalinization plants use enormous quantities of energy and also produce saline brines as waste products. The energy intensity of desalinization processes often represents the single most significant barrier to wide-scale adoption of the technology. Between the energy consumed in the process and the waste produced, proposals for the construction of desalinization plants require careful analysis and balancing of water, energy, environmental factors, and economics.

DESERT PUPFISH

It might seem ironic that several fish species have their name associated with the word *desert*. Still, in this instance, their hardiness and adaptation to desert hydrology are what make these novel creatures inhabitants of the region. Named for their resemblance to frolicking puppies, the desert pupfish (*Cyprinodon*) are members of several species and subspecies of rare bony fish that nip at

Figure 2.65. Desert pupfish (source: https://en.wikipedia.org/wiki/Desert_pupfish; public domain image).

the tails of prospective mates as they dart back and forth in desert pools and rivers. These fish are remarkable for their persistence and endurance in springs, marshes, ponds, and creeks of southern California, southwestern Arizona, and northwestern Mexico. Here in these desert regions, water temperatures can exceed 110 degrees Fahrenheit, yet these fish survive and even thrive. Their reputation for endurance in extreme environments is assured when you consider that some of these same desert water holes have salinities that are nearly double that of seawater. You can see desert pupfish in the Ash Meadows National Wildlife Refuge, Death Valley National Park, California's Anza Borrego Desert State Park, and in the Amargosa River of California and Nevada.

Despite their hardiness, threats to desert pupfish come from a host of impacts. These impacts include water diversions, competition from nonnative species, habitat degradation from overgrazing, and agricultural activities. The fish underwent rapid declines in the years after 1950. Desert pupfish survive in only eleven known native locations, along with several transplanted sites across their range. Although the species populations appear stable at present, the small number of places where they live means that they are vulnerable to random events that can harm individual pools and from regional changes such as continued desertification and global warming.

Figure 2.66. The Gila River, a formerly perennial river, once supported Indigenous communities all the way from its source in New Mexico to its confluence with the Colorado River at Yuma, Arizona. Except for the occasional flash flood, modern water development has so thoroughly dewatered the river in its lower reaches that, as shown here, road departments no longer bother to build bridges across the riverbed. The railroad bridge seen in the background was built before the Gila River was completely dewatered. Photo courtesy of the author.

DEWATERED STREAMS

One of the sad legacies of water development in the American West is dewatered streams. Under the Prior Appropriation Doctrine, water users may claim the rights to water from naturally flowing rivers and streams. In the nineteenth century, when colonists formulated this doctrine, no provisions for protecting rivers existed. What this means is that water developers could, and did, appropriate and divert every drop of water flowing down a creek. Their water diversions left thousands of rivers severely depleted and often entirely devoid of water below diversion dams. It wasn't until environmental awareness expanded in the late 1960s and 1970s that people began taking note of these depleted streams. During this era, states first recognized that water in streams constituted a beneficial use for water, and began taking the administrative steps necessary to restore waterways. Since then, many states have developed instream flow programs to get water back into depleted waterways, but because the rights to diversions are so senior, the cost and logistics remain daunting. For this reason, even to this day, many western streams remain severely depleted or totally dry below major diversion structures.

THE DITCH

In the eastern United States, people use ditches to drain water from fields. In the West, growing crops requires supplemental irrigation, so people use ditches to transport water from rivers or reservoirs to fields. Indigenous farmers built

Figure 2.67. This is the San Luis People's Ditch, the oldest continuously used ditch in Colorado. It dates from April 10, 1852, and holds the number one priority water right in Colorado. Photo courtesy of the author.

the earliest known ditches in the West. In Arizona, the Hohokam people began building ditches by at least 100 AD and the Ancestal Puebloans by about 900 AD. In southern Arizona and northern Sonora, some ditches in the Early Agricultural Period date as far back as 2000–2500 BC. When the Spanish arrived in the 1500s, they brought ditch building technology with them. In short order, the Spanish incorporated Indigenous knowledge into their practices in ways that gave rise to the acequias that exist throughout northern New Mexico today. With the arrival of the Americans, they too began constructing ditch systems that now crisscross much of the American West. By the early 1870s, real estate speculators and developers realized that they could buy land and run a ditch to it, and thereby drastically increase the value of the formerly dry land. Entrepreneurs initiated a flurry of ditch-digging and land speculating that continued more or less unabated until the 1920s.

Typically, each ditch includes several infrastructure components. These components include a diversion dam, a headgate to let water into the ditch, sluice gates for removing sediment, and a gage for measuring flow. Other components include flumes and siphons for going over or under landscape obstacles, laterals for delivering water to individual users, and sometimes fish passages or kayak chutes to let fish and people around the diversion dams.

DITCH EASEMENT

I suspect that ditch easement conflicts have put more children of attorneys through college than any other water matter. These conflicts arise because the seemingly opaque rights of ditch companies to manage water flow in easements across private property can clash with the perceived rights of landowners along the ditch. Often the results are bitter lawsuits that yield hefty legal fees for all the attorneys involved. Take a look at a ditch, and you'll notice that as it winds across the landscape, it often crosses private property. An easement provides legal access across the property not owned by the ditch company. Ditch companies usually hold a bundle of rights associated with ditch easements. Courts across the West have recognized that ditch companies can inspect, operate, maintain, and repair their ditch. Courts in Colorado, for example, have held that the owner of a ditch easement "may do whatever is reasonably necessary to permit full use and enjoyment of the easement."[113]

In this instance, it sometimes comes as a surprise to nearby landowners that the ditch company has the right of "ingress and egress for maintenance, operation, and repair."[114] As one might surmise, a landowner and a ditch company's

Figure 2.68. This is a typical ditch cleaning project in a residential neighborhood in Boulder, CO. Photo courtesy of the author.

understanding of what is "reasonably necessary" can be quite different. Digging ditches often established what attorneys call *prescriptive easements*. These easements develop over extended historical use. It's analogous to adverse possession, but the ditch company only has the right to move water and do what it takes to maintain the ditch. Managing prescriptive easements for ditches is often somewhat complicated. From the perspective of the water user or the landowner, many potential points of disagreement may arise. Landowners often object to mechanical equipment crossing land that they thought was their private estate. Ditch companies object to impediments placed or trees planted within the proscriptive easement by the landowner. In other words, easement management priorities shift depending on your perspective.[115]

DOMINY, FLOYD ELGIN

> "[The] former commissioner of the Bureau of Reclamation should be required to dismantle the dam with hammer and star drill. John Wesley Powell admired, perhaps revered, what he saw in Glen Canyon. Floyd Dominy rejoiced in destroying it."—David Brower[116]

Figure 2.69. One of the great mandarins of western water development, Floyd Dominy served as the Commissioner for the US Bureau of Reclamation from 1959 to 1969. Image courtesy of the University of Wyoming, American Heritage Center, Floyd E. Dominy Papers, Accession Number 02129-1981-05-26, Box 8A, Folder 3.

"Floyd E. Dominy, a child of the Dust Bowl who pursued his dream of improving nature and human society by building vast water projects in the West—steamrolling over pristine canyons, doubtful politicians and irate conservationists—died on April 20 in Boyce, VA." This was how the *New York Times* began Dominy's 2010 obituary of the man who, as the Commissioner of the Bureau of Reclamation, led the construction of the Glen Canyon, Flaming Gorge, and Navajo Dams in the Colorado River Basin. Dominy's Bureau of Reclamation presided over building the Trinity Dam as part of California's Central Valley Project and failed in attempts to place dams in Dinosaur National Monument's Echo Park and within the Grand Canyon National Park.[117]

To many, Dominy's name will forever remain associated with his nemesis, David Brower of the Sierra Club, who led public relations efforts to stymie Bureau of Reclamation dam projects around the West. Dominy's disdain for Brower was legend. At a conference of water users, I heard Dominy state to the ecstatic crowd that he had once called Brower a "sanctimonious ass," adding that he regretted calling Brower "sanctimonious." Dominy was never one who shied away from self-promotion or taking credit: In the same speech that he spoke of his relationship with Brower, Dominy boastfully described building the western power grid in the first person. And his claims were not far from the truth. Beginning as the assistant commissioner in 1957, associate commissioner in 1958, and commissioner from 1959 to 1969, his agency vastly expanded the amount of water it impounded, much of it for irrigation, municipal supply, and electric production.[118]

DOWSING (WATER WITCHING)

Hydrologists lump dowsing or water witching into the dustbin of pseudoscience. The US Geological Survey describes water dowsing as the "practice of using a forked stick, rod, pendulum, or similar device to locate underground water, minerals, or other hidden or lost substances."[119] As the US Geological Survey points out, in many regions "of adequate rainfall and favorable geology, it is often very difficult *not* to drill and find water!"[120]

Figure 2.70. Otto Elder von Graeve (1872–1948), shown here holding a divining rod, was considered an expert on water dowsing. Image courtesy of the Library of Congress. Public domain image.

Even so, I think it is worth departing from the format of this book by telling a personal story about water dowsing. Back in the late 1980s, I was working as a hydrologist for the Denver Water Board. One of my projects was to develop groundwater supplies for a golf course in South Denver. As part of the project, we needed to drill some test holes to determine the aquifer characteristics. Before we could dig the test holes, we called in utility locators to make sure we did not drill into anything. It turns out that there was a large water main in the area. When the utility locators from the water department came by the site, they explained that because the main was so old (there were no wires attached to the pipe that they could electrify and find the pipe), they would have to locate it through dowsing. The locator took two copper wires he previously bent into an "L" shape. Holding each wire loosely and parallel from each other from the short end, he walked out across the field. Some distance out, the wires began moving together and then crossed, and as he continued, they uncrossed. The locator pronounced the water main under the spot where the wires crossed.

Ever skeptical, I figured that the locator was playing a trick on the young hydrologist. So, I grabbed the copper wires from his hand and marched across the field. As I advanced, the wires crossed and then uncrossed at the same spot just as they had done before. To this day, I can't fully explain what I saw and experienced. But I cannot deny what I saw. My best guess is that flowing water creates a slight current. Water, after all, is a polar molecule and is therefore slightly magnetic. And, as they teach you in high school physics class, moving magnets create a current. Perhaps because I was walking with two metallic rods, and because the water was flowing in the water main, the current induced a magnetic response in the rods, causing them first to draw together and then move away as I advanced. Until a researcher can propose a scientifically sound hypothesis that explains the mechanism behind dowsing that hydrologists can test and reliably reproduce, the practice will remain discredited among professional water managers.[121]

Despite my single experience, I still think it is best to use a groundwater hydrologist to identify the site to dig a water well. By the way, when we then drilled the test holes, none of them struck the water main.

DRIP IRRIGATION

Attempts to carefully control the amount and distribution of water onto agricultural fields have been ongoing since ancient times. But it was only in the late 1950s and early 1960s when Israelis began to experiment with plastic tubing and specialized emitters that technology of modern drip irrigation came

Figure 2.71. In many agricultural districts, scarce water supplies have led farmers to install efficient irrigation systems. Here, Flavio Chegue, an employee of Mameda Farms, picks cantaloupes near Rocky Ford, CO, in 2005 in a field that deploys a drip irrigation system. Some municipalities, such as Aurora, CO, which owns a large stake of water rights in this agricultural area, see it in their political interest to preserve farming by helping install drip irrigation systems. These systems can use up to 1/3 less water than flood-irrigated fields. Source: Photograph by Maria J. Avila, Rocky Mountain News. Denver Public Library, Special Collections.

into being.[122] Because of water scarcity and the need for food security, Israel was at the forefront of developing highly efficient water distribution systems. The Israelis were the first to patent drip irrigation systems and wasted no time exporting the technology abroad. At first, drip irrigation was deployed primarily in arid regions, but farmers began using it in any area where the careful application of water would enhance crop growth.

Essentially, drip irrigation is a form of micro-irrigation that utilizes plastic pipes and flow release valves (known as emitters) to deliver water and nutrients directly and slowly to plants roots. Drip irrigation also benefits from the use of low-pressure water flows that reduce evaporation and runoff.

DROUGHT

In the era before people could readily transport commodity food products from one region or country to another, drought often resulted in a slow death for

Figure 2.72. Arthur Rothstein's photograph of overgrazed, drought-parched land in Pennington County, SD, made in 1936, was influential in shaping New Deal–era policy toward Dust Bowl recovery. Controversy followed this image with allegations that Rothstein staged it by placing the skull to heighten its emotional impact. However, Rothstein admitted that he indeed moved the skull a few feet but asserted that did nothing to diminish the documentary value of the photograph. Moreover, Rothstein maintained that he did no wrong. Nevertheless, Rothstein's photograph became an enduring visual trope that influenced the public's image of drought. (Courtesy of the Library of Congress, image LC-USF34-004389, public domain)

multitudes of people through starvation and thirst. In developed countries today, drought can be severe and devastating but seldom results in the death of the locals living through the tragedy. It's helpful to think of drought as a dynamic concept. As the US Drought Monitor explains, how you recognize drought depends on how it affects you.[123] Drought contains quantitative elements such as reduced rainfall, low soil moisture, and diminished streamflow. Also, qualitative concepts including climate variability and human factors such as population size and water use affect drought severity. Unlike a lightning strike or a flood, drought is a chronic problem whose consequences intensify over time. Generally, hydrologists consider drought as a lowering of precipitation or streamflow compared to long-term averages, or a prolonged abnormal moisture deficiency. For humans, the impact of drought is dependent upon the many variables mentioned above, plus the society's ability to shift resources around to mitigate its effects. Along with that, people's values toward conservation as well as attitudes toward community versus private good, pricing issues, and local agriculture considerations affect how the nation responds to drought.

It is also important to note a distinction between drought and more pernicious drying events: In mega droughts, the dry conditions last two or more decades. Beyond that comes aridification, in which the drying becomes an essentially permanent climatic condition.

Several federal and state entities maintain the US Drought Monitor to facilitate federal, state, local, tribal, and basin-level decision-making to respond to drought. Essentially, the US Drought Monitor synthesizes various drought severity measurements to produce standardized maps of drought severity around the United States.[124] You often see these maps reproduced in newspapers and magazines that contain articles about drought. On the maps, one will find a sliding scale of severity ranging from abnormally dry, moderate drought, severe drought, extreme drought, and, finally, exceptional drought. At the abnormally dry level, impacts might include short-term dryness, slow planting, and reduced growth of crops or pastures. These impacts gradually intensify until exceptional drought sets in, where we see unprecedented and widespread crop/pasture losses and shortages of water in reservoirs, streams, and wells that create water emergencies. Although the US Drought Monitor does not reveal why a drought occurs, it is valuable for comparing drought conditions between regions and over long periods.

DUST STORMS

In the aftermath of the Indian Wars of the 1860s and 1870s, Euro-Americans rushed onto the High Plains to begin establishing farms. Many of these farmers began plowing up the tough prairie sod to plant commodity crops such as wheat, barley, and oats. With the expansion of railroads into the region, even the most remote areas of the plains were soon functionally closed to eastern markets. Colonists quickly exploited the High Plains to grow commodity grains. From the 1880s on, grain production through dryland farming steadily expanded. Then, with the onset of World War I, commodity prices soared, and farmers dramatically expanded output to cash in on the opportunities that high grain prices entailed. Grain monocultures replaced formerly diverse prairie grassland ecosystems characterized by dense clusters of native grasses. Once the war ended, commodity prices began declining and then collapsed. By the start of the Great Depression, formerly productive farms lay fallow, exposing fragile prairie soils.

Figure 2.73. In 1936, during the height of the Great Depression, dust storms raged across the High Plains. Commodity booms in wheat and other grain led farmers to plow up vast areas of prairie. When natural cycles of wet weather gave way to drought, the destabilized topsoil began to blow under dry, windy conditions. Here is one such storm in Baca County, CO. Source: Photograph by D.L Kernodle. Library of Congress Call No. LC-USF34-001615-ZE. Public domain image.

Nonirrigated lands were particularly vulnerable to soil erosion. Calamity eventually struck as deep drought set in, killing off vast areas of crops. Then as dry winds began to blow, vast dark brown clouds of dust, sometimes called *black blizzards* or *black rollers*, swept across the country. In the worst-hit areas, the dust storms were so thick that victims' lungs filled with fine silt, inflaming their alveoli to induce a kind of dust borne pneumonia.[125]

Dozens of dust storms struck in the 1930s, affecting some 100,000,000 acres from Texas to Canada. Newspapers referred to the region as the Dust Bowl. Among the worst storms was one that struck on April 14, 1935, known as "Black Sunday." It swept the entire Great Plains, from Canada south to Texas. With dust reaching New York and Washington, the federal government stepped in to provide relief and to enact reforms that would eventually stabilize the situation. The storms resulted in massive human migrations out of the Great Plains, immortalized by epic works such as John Steinbeck's *The Grapes of Wrath*.[126]

ECHO PARK CONTROVERSY

One of the most bitter environmental battles of the twentieth century took place over a remote corner of northwestern Colorado and northeastern Utah. Here, the Green and Yampa Rivers flow through spectacular red-rock canyons

Figure 2.74. A photograph of Steamboat Rock with depictions of the water depth for the proposed Echo Park Dam. This image is part of an unattributed informational poster from approximately 1955.

containing geological formations that span some 1.2 billion years of earth history. John Wesley Powell named many of the most prominent features on his 1869 expedition, including the Gates of Lodore, Whirlpool Canyon, Split Mountain, and Echo Park. President Woodrow Wilson established Dinosaur National Monument in 1915 through an executive order. Then, in 1938, President Franklin Roosevelt enlarged the monument to protect its remarkable geological, paleontological, and archaeological resources. But with the passage of the 1922

Colorado River Compact and the 1948 Upper Colorado River Basin Compact, water developers began advocating for dams across the region, including a massive structure at the heart of Dinosaur.

By the mid-1950s, forces for development and conservation lined up for an angry fight that would eventually derail the big dam-building era in America. Politicians, including Wayne Aspinall and Bureau of Reclamation Commissioner Floyd Dominy, became tireless advocates for flooding the monument. Pitted against them were conservationists, including David Brower, historian Bernard DeVoto, writer Wallace Stegner, rafter Martin Litton, and many other concerned citizens. For a time, the dam proponents had the upper hand, going so far as to win support from Douglas McKay, President Dwight Eisenhower's Secretary of the Interior. But the conservationists pushed back hard, and the Sierra Club, led by David Brower, began taking out full-page ads in the *New York Times* and *Los Angeles Times* opposing legislation for the project. In time the bill stalled. Eventually, Brower cut a deal that removed the Echo Park Dam from the authorizing law but allowed dams at Flaming Gorge and Glen Canyon to proceed. Environmentalists had saved Echo Park. But it was a Pyrrhic victory. Brower had never visited Glen Canyon, and once he did in the days during the final construction of the Glen Canyon Dam, he realized that he had made a grave error by trading off one natural wonder for another. Brower would go to his grave lamenting his compromise.[127]

EFFLUENT

Water users consider an *effluent* as any wastewater discharged into a river, lake, aquifer, or the sea. This term does not distinguish the wide variety of wastes emanating from nonpoint source pollutants at sites as variable as parking lots to industrial-scale discharges, including agricultural feedlots, municipalities, and industries. It does include all kinds of wastes from treated municipal sewerage to discharges of hot water from power plants, and chemical residues dumped illegally in the night. In recent years, there have even been battles over whether the releases from reservoirs, due to altered water temperatures, should be classified as effluent and managed under the Clean Water Act.

In the United States, the Clean Water Act regulates the discharge of effluent. There are major financial incentives to cheat in cleaning and reporting discharges of effluent due to the costs of managing pollutants. Of course, many dischargers are responsible and manage their waste correctly, but the few that do not cause considerable damage to our waterways. Economists call the discharge of effluents

Figure 2.75. Seen here is the effluent outfall from the Denver Metro Waste Water Reclamation District wastewater treatment plant in North Denver. When I took this photo (March 18, 2015), upstream diversions had reduced the South Platte River at this site to just 8.3 cfs, while the combined flow of the two outlets was about 254 cfs. In other words, more than 30 times more effluent was present at this site than native streamflow! Water in this waste stream includes much of the treated sewerage gathered from across the Metro Denver area. While regulators say that the water is safe for body contact, one look at the water (let alone the odor that was present) suggests that anyone considering jumping in at this site has got to be mad. Photo courtesy of the author.

that affect downstream water users to be "third-party impacts," as polluters pass on the cost to deal with impacts to downstream users. In the West, where the climate is generally more arid than other regions, the management of effluent takes on greater importance because the rivers and streams in this area have lower average discharges and, therefore, lesser ability to absorb and diffuse wastes.

ELEPHANT BUTTE DAM

At the time of its completion in 1916, journalists heralded Elephant Butte Dam as the largest irrigation dam in the world. Spanning the Rio Grande seventy-five miles north of El Paso, Texas, the solid masonry dam is approximately 1,675 feet long and can impound nearly 2.2 million acre-feet of water to a depth of almost 200 feet.[128] Built by the US Reclamation Service (today's Bureau of Reclamation),

Figure 2.76. Here is a panoramic composite view of the Elephant Butte Dam and Reservoir made in 2016. Photo courtesy of the author.

the project transformed the desert of southern New Mexico to became one of the early successes of federal reclamation. Water from the project is distributed from Sierra County, New Mexico, south to El Paso and into Mexico.

Proposals for a dam on the Rio Grande began circulating in the 1880s. The impetus for a dam gained traction as irrigation expanded dramatically in southern Colorado in the 1890s. Colorado's increased diversions also coincided with severe drought conditions in the Rio Grande Basin. These events led to diminished flows on the Rio Grande in southern New Mexico. Water users soon proposed various public and private reservoir projects, including one funded by English capital. Eventually, the site at Elephant Butte emerged as the preferred option. However, the Reclamation Service did not construct Elephant Butte without controversy. One lobby, the International Dam Commission, an assembly of Mexican and American engineers, wanted the dam to straddle the US-Mexican border. Their proposal would have allowed Mexico to own half of the dam, reservoir, and water supply. The Reclamation Service preferred a site north of El Paso within New Mexico, which allowed the United States to control the operations of the reservoir fully. Congress sided with the Reclamation Service and authorized the construction of the dam at its present location in New Mexico. But its construction touched off an international debate over the allocation of water between Mexico and the United States, a discussion that was later replicated on other transnational rivers in the West throughout the twentieth century.[129]

ELWHA DAM

Some of the most productive salmon runs in the Northwest once took place on Washington's Elwha River as fish migrated to their spawning grounds in what is

Figure 2.77. The Elwha Dam removal stands as a symbol for riparian renewal. Here is the Elwha Dam during its deconstruction on October 20, 2011. Image courtesy of the National Park Service. Public domain image, source: https://www.nps.gov/olym/learn/nature/dam-removal.htm.

now Olympic National Park. When hydropower development came into vogue during the late nineteenth and early twentieth centuries, engineers targeted the Elwha and its deep canyon as a logical site for hydroelectric dams. Promoted by the entrepreneur Thomas Alwell, the Elwha dam was constructed starting in 1910 and was functioning by 1913. A second dam was built in 1927 eight miles upstream within Grimes Canyon to increase the power output from the watershed. With the construction of these dams, the salmon fishery soon collapsed.

Eventually, Congress created Olympic National Park in 1938, even though two hydropower dams spanned the Elwha River deep within the park. With the rise of the environmental movement, possible restoration of the Elwha Watershed to provide for salmon migration became hotly debated in the Northwest. After many years of politicking and negotiations, environmentalists secured approval for deconstruction and removal of the Elwha and Glines Canyon Dams. Work to raze the dams began in 2011. These became the largest dam removal projects in history and demonstrated that it is, in fact, possible to restore watersheds impacted by old dams.[130]

ENDANGERED SPECIES ACT

Among the most powerful of all the American laws to protect the environment, the Endangered Species Act (ESA) was signed into law by Richard Nixon in 1973.[131] The purpose of the ESA is to protect and recover imperiled species and the ecosystems upon which they depend. Under the ESA, the US Fish and Wildlife Service may list species as either endangered or threatened. *Endangered* means a species is in danger of extinction throughout all or a significant portion of its range. *Threatened* means a species is likely to become endangered within the foreseeable future. All species of plants and animals, except pest insects, are eligible for listing under the act. In the ESA, Congress defined species to include subspecies, varieties, and, for vertebrates, distinct population segments.[132]

Because protections for species are so broad, an ESA listing can affect water and land use management decisions across whole watersheds. Thus, if a species needs water for part of its life cycle, even if the water is only to support some aspect of its habitat, federal decisions must take that information into account. Because of this, water management decisions will require specific quantities of water to go toward the environmental needs of that species' recovery goals. For

Figure 2.78. The greenback cutthroat trout (*Oncorhynchus clarkii stomias*) is a fish listed as threatened under the Endangered Species Act. Because of its status as a threatened species, state and federal agencies have strived to protect its habitat and recover its numbers. If it were not for the Endangered Species Act, the greenback cutthroat trout would likely be extinct. Photograph Bruce Roselund, US Fish and Wildlife Service. Public domain image.

this reason, dam releases on the Colorado River have been modified to benefit endangered fish, flows on the Platte River have been modified to enhance whooping crane habitat, and dam operations in the Northwest have changed to help salmon needs. Antienvironmental lobbies and the politicians that support them have worked to undermine the ESA almost from the moment that Congress enacted it. Conversely, environmental groups have routinely entered court cases to influence land and water use decisions to protect species, citing the authority of the ESA.

ENVIRONMENTAL MOVEMENT

Environmental activism has deep roots in the United States that trace back to nineteenth-century intellectuals such as Walt Whitman as well as artists including Alfred Bierstadt and Thomas Moran. By depicting American landscapes in sublime terms, they set the stage for the protection of untrammeled landscapes and the creation of national parks. As a movement, American environmentalism congealed around the ideas of John Muir, the avid hiker and environmental philosopher who founded the Sierra Club in 1892. Muir is distinguished for his efforts to conserve wild nature and his opposition to the view of nature as something that Americans could alter whenever there is a human benefit. The American environmental movement had its first major battle, and perhaps most famous defeat, in its failed attempt to foil the construction of the Hetch Hetchy dam within Yosemite National Park. Muir, who led the fight against the dam, died heartbroken with the submergence of the valley that he believed was every bit as beautiful as the Yosemite.

Despite Muir's defeat, the movement continued to contest reservoir projects in the American West, culminating in the Pyrrhic victory by Muir's intellectual and administrative heir, Sierra Club president David Brower. Through Brower's political leadership and tenacity, the conservation community forced the US Bureau of Reclamation to scuttle plans for a dam in Echo Park, at the center of Dinosaur National Monument. In exchange for abandoning the dam at Echo Park, Brower agreed to a dam at Glen Canyon—a place he had not yet seen—a concession that he would regret the remainder of his life.

Through the 1960s and 1970s, the environmental movement expanded its scope and broadened its interests as public concern about the environment ballooned in the wake of the first Earth Day. In about a decade, the United States passed into law the Wilderness Act (1964), Clean Air Act (1970), National Environmental Policy Act (1970), Clean Water Act (1972), Endangered Species

Figure 2.79. With Edward Abbey's publication of *The Monkey Wrench Gang*, the idea of destroying or deconstructing the most damaging dams entered the popular public imagination. Here on the Matilija Dam in Southern California, a structure that the environmental community seeks to remove, an anonymous artist translated that idea into a landscape-sized editorial. Image courtesy of California Trout. Photograph by Michael Wier. Used with permission.

Act (1973), and other signature laws that changed America's relationship to the environment. Although the American public has dramatically benefited from significant strides in environmental quality with the passage of these laws, these successes have led to complacency as the most egregious environmental abuses have receded into memory. With Americans forgetting the steep costs of polluted skies and filthy water, the environmental movement often finds itself fighting losing battles. Opposition to environmental regulations comes from conservative politicians and their corporate backers who lambast environmental laws as job killers that constrain business profits. In the twenty-first century, the environmental movement is engaged with issues including global warming and climate change and related concerns about ocean resources, environmental justice, deforestation, and freshwater protection. Compounding these challenges are the growing rejection of scientifically based policy making within the general public and disinformation campaigns by special interests who wish to steer public land and resources into private hands.[133]

EVAPOTRANSPIRATION

In arid states where water rights are valuable, the careful measurement of the total amount of water users consume—the consumptive use—is critical in determining the outcome of water rights court cases. Central to this is determining the amount of water delivered to plants and how much they consume. These measurements have tremendous economic consequences for all involved.

Engineers define *evapotranspiration* as the sum of evaporation and leaf transpiration at any given site. Although simple in concept, practical measurement of evapotranspiration is a daunting task. Hydrologists employ devices known as lysimeters to carefully measure the amount of water plants use. They accomplish this by excavating a block of soil, taking pains to prevent it from cracking and crumbling, and encase it in a box that is open to the air. By recording the variations in the weight of the box over time, they can estimate the changes in soil moisture. When engineers link this information to a weather station, they can

Figure 2.80. Allan Andales of the Colorado State University Department of Soil and Crop Sciences standing beside an undisturbed block of soil known as a lysimeter monolith during its installation in 2008. Workers lifted the monolith using a crane and placed it on a weighing scale enclosed in an underground chamber. By recording the variations in the weight of the box over time, technicians can estimate changes in soil moisture. Photo by Lane Simmons, courtesy of Colorado State University. Image source: http://vpr.colostate.edu/pages/climate_change_cloudsat.html.

determine evapotranspiration rates for different crops. Evapotranspiration data derived from lysimeter studies form the basis of many water rights change cases. The higher the estimates for evapotranspiration, the more water a water right might represent. Consequently, vast sums of money are spent by applicants and their opposition to the proposed water right changes by contesting every detail in the measurement of crop consumptive use.[134]

FISH HATCHERIES

Since the earliest days of Western settlement, Euro-Americans began building ponds and stocking them with fish. For many, these fish provided necessary protein for themselves, and if there were enough, they could sell these fish to hungry miners or travelers. This is how the common carp ended up in many western lakes and waterways. Even though we rarely eat carp today, in the nineteenth

Figure 2.81. Almost from the beginning of Euro-American settlement in the American West, colonists started importing and stocking fish into Western creeks and rivers. As settlement expanded, federal and state agencies constructed fish hatcheries in every state to promote recreational fishing. Today, the mission of many fish hatcheries has diversified and now includes efforts to recover threatened and endangered species. Shown here are wild American paddlefish (*Polyodon spathula*) captured from the Missouri River in South Dakota and transported to the Gavins Point National Fish Hatchery, Yankton, SD, for spawning. Biologists will hold the fish for the three-day process and then return them to the Missouri. Photograph by Sam Stukel, courtesy of the US Fish and Wildlife Service. Public domain image.

century these hardy fish were easy to raise and added an important supplement to a colonist's meager fare.[135]

From those modest beginnings of raising fish for protein, the modern hatchery industry evolved. Today, there are over thirty fish hatcheries operated by the US Fish and Wildlife Service across the West. Each state also manages a significant number of hatcheries to support the local fishing industry and to assist fishery conservation efforts. In Idaho, for instance, nineteen state-operated hatcheries raise and support multiple species, including both sport fishes and species of concern. In addition to governmental agencies, many more private hatcheries and ponds supply fish for private clients and fishing groups.

Fish hatcheries have a mixed record for environmental stewardship. On the one hand, they are essential to efforts to preserve many native species populations and to avert the extinction of rare endemic fishes. On the other hand, fish hatcheries have contributed to the transformation of western waterways through the deliberate, systematic introduction of nonnative sport and predator species and unintentional introduction of fish diseases and aquatic pests such as whirling disease and the New Zealand mudsnail.

FISHING

Fish are healthy and tasty sources of protein and nourishment. It's not surprising then that the Indigenous people wasted no time utilizing fish from the moment they entered North America. In interior rivers, the evidence for early fishing is scarcer, but when archaeologists find it, it is indisputable. Fish hooks, images on Puebloan pottery, and rock art images all point to continuous use in the past throughout the West.

European and later American explorers and migrants to the West wasted no time exploiting the fisheries of western rivers. They actively began Europeanizing the western rivers through the deliberate introduction of fish species native to the Old World. As we have seen elsewhere, vast numbers of carp now found in western lakes and rivers are descendants of fish originally stocked during Gold Rush days by colonists seeking reliable sources of protein. Likewise, trout, Tiger Muskie, and other species were introduced to western rivers to enhance recreational fisheries beginning in the nineteenth century. High-mountain lakes, many that had never before contained fish, were stocked to support backcountry recreational fishing. Eventually, state and federal agencies began programs to eradicate endemic fish to enhance the sport fishery (primarily by introducing European trout) in many western rivers. Perhaps the most infamous program was the systematic dumping

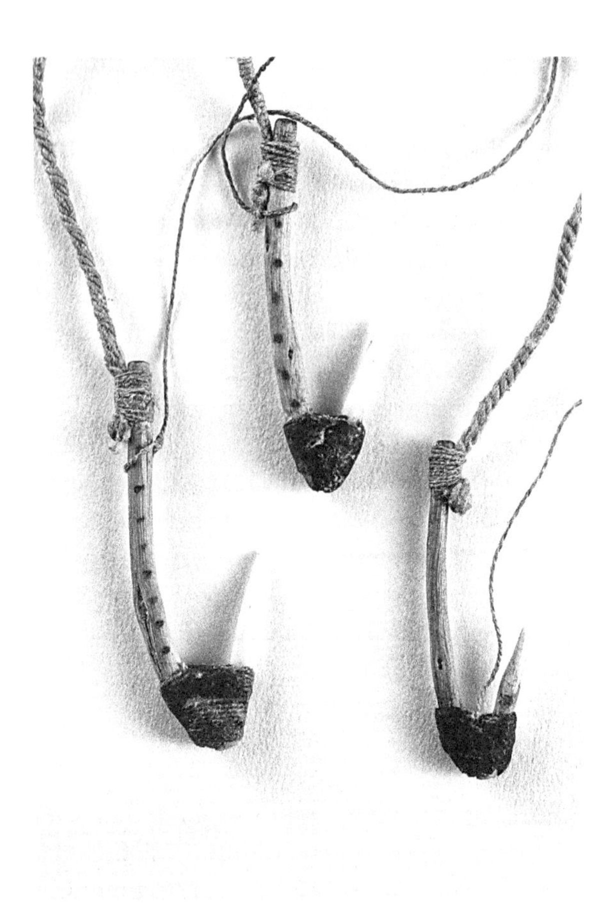

Figure 2.82. These delicate but exceptionally well made fishhooks found in a Fremont culture excavation dating from about 600 AD point to extensive use of fish in Indigenous diets on the Colorado Plateau. Photograph by Francois Gohier. Copyright University of Colorado Museum of Natural History UCM05960.

of *rotenone*, a notorious fish poison, by the US Fish and Wildlife Service in the early 1960s to eliminate endemic "trash fish" in the Colorado, Green, and San Juan Rivers. Their goal at the time was to stock the river with sport fish. Amenity fish, as the thinking went, would bring in dollars from sportsmen. In due time, the native giants like the pikeminnow and other endemic fish became endangered, and now no one is sure if we can even bring them back. There is a certain dark irony in this, as the US Fish and Wildlife Service is now committed to spending large sums of money to correct its mistakes.

FLASH FLOOD

Sudden localized flooding, typically due to a heavy rain event from intense slow-moving thunderstorms, is referred to as a *flash flood*. When rain falls so fast that the soil cannot absorb it quickly enough, water moves overland and

Figure 2.83. During this flash flood in New Mexico during 1979, this man became stranded atop his vehicle. Photograph by Dennis Dahl. Image courtesy of the New Mexico History Museum, photographer Dennis Dahl, Santa Fe New Mexican archive, Negative Number HP.2014.14.366.

down drainages, mounting in intensity as the quantity of water increases. Flash floods are extremely destructive and violent and often sweep away everything in their path. In desert canyon country, thunderstorms may occur some distance away, but water can rip down affected canyons increasing the flow from a trickle to a raging torrent in a matter of minutes. If the stream channel is bedrock or contains other relatively impervious materials (such as pavement or compacted clayey soils), it becomes difficult for the ground to absorb the water. It results in high amplitude, short-duration flow events. Likewise, if it has been raining or wet for some time and the soils become saturated, even porous soils cannot keep up with infiltration. If at that point a cloudburst occurs, the added water overwhelms the watershed, and a flash flood may ensue.

FLOODPLAIN

In the long list of water resource terms that seem poorly defined, few are as loosey-goosey as the term floodplain. The US Geological Survey describes floodplains as the strip of relatively flat and normally dry land alongside a stream, river, or lake that becomes covered by water during a flood.[136] That seems simple

Figure 2.84. These side-by-side photographs show the flood plain of Clear Creek, near present-day Georgetown, CO. On the left is a view from the Empire Trail in 1873, and on the right is the view from the same location in 2015. Today, Interstate 70 occupies as much as twenty percent of the former floodplain. Clear Creek's flood plain mostly includes the riparian vegetation along the river. In the West, people developed or encroached upon many floodplains like this one. Roads, railroads, gravel pits, urbanization, farming, and recreational trails, to name a few, have altered the condition of floodplains and changed their ability to convey floods and provide ecological services.

enough, but do you draw the flood plain boundary at the edge of a one-year, ten-year, one-hundred-year flood, or some other magnitude event? What about developed areas in bottomlands that would flood but now have levees or other structures separating them from the water? Additionally, some floodplains might no longer qualify as such because upstream dams or diversions change downstream hydrology.

My point in raising these questions is that unless you carefully describe the flood size and what features to include or exclude, determining the *extent* of any particular floodplain is very difficult and subject to many judgement calls.[137] Rivers contain an active floodway and, beyond that, a flood fringe.[138] These are inherently "fuzzy" terms that involve statistical analysis and probability mapping. Hydrologists point out that many people interchangeably use the terms floodplain, flood zone, flood prone area, active floodplain, and riparian zone. This is problematic because these terms actually describe slightly different locations on the landscape. Compounding matters is that only occasionally will all these locations coincide with each other.[139] Overall, when dealing with floodplains, one needs to keep in mind that these are highly variable systems.

FLOODS

Many floods cause dramatic death and destruction that make grim headlines in newspapers and on the evening news. Floods might seem unpredictable, but hydrologists, engineers, and planners have made great strides in predicting and mitigating these events' severity. Hydrologists regard any high streamflow event that overtops a natural or human-made stream bank to be a flood. Once the water overtops the bank, it spreads out to inundate the flood plain. Floods and their effects vary dramatically within and between streams. One often hears of five-year, fifty-year, one-hundred-year, and so on flood events. These intervals refer to the statistical probability of a flood event occurring in any given year. A one-hundred-year flood event, for example, has a 1 percent probability of occurring in any given year. Hydrologists analyze many years of streamflow data to estimate the size of various floods. The more data that is available, the more accurate the flood analysis becomes. A common mistake many people make is that when they hear that a hundred-year flood has occurred, they believe the event will not repeat for another one hundred years. This understanding cannot be further from the truth. Instead, a one-hundred-year magnitude flood may happen many years in a row. However, over a very long period, the likelihood is that it will occur only in one percent of the years.

Figure 2.85. The Black Hills flood that hit Rapid City, SD, is among the worst floods in US history. On June 9–10, 1972, extremely heavy rain fell on the eastern Black Hills of South Dakota. This produced record floods on Rapid Creek and other nearby streams. Approximately 15 inches of rain fell in 6 hours near Nemo, SD, and a sixty-square-mile area saw more than 10 inches of rain. The resulting flooding left 238 people dead and 3,057 people injured. Property damage exceeded $160 million. Photo courtesy of the Rapid City Public Library.

Additionally, consider that channels vary widely in size and shape. Small or large, each stream or river tends to flood proportionally to its size. A small creek that is flooding can still kill people or wreak local havoc, whereas a massive flood in a major river can cause widespread devastation and death and make national or international news.

Floods might arise from purely natural events (thunderstorms, hurricanes, rapid snowmelt) or might be wholly human-made (dam breaches) in origin. Flood impacts, however, tend to be either compounded or mitigated by natural conditions and the built environment. Gilbert White, the geomorphologist best known for his work in natural hazards, has shown that the effects of floods are often greatly intensified because of land-use decisions in flood zones. A landowner might gripe when a governmental entity restricts their ability to build in a flood zone. Conversely, in an area of lax land-use planning, the people who buy houses in flood-prone zones often seek governmental assistance once a flood destroys their property. As White has written, it's ultimately best to avoid building in these zones rather than burden society with recovery costs. That said, people have historically built in flood-prone areas. This development creates political challenges for government agencies, especially when property owners have unwittingly moved into flood zones.[140]

Figure 2.86. This photograph from 1890 shows the Montrose Placer Mining Company flume, in Montrose County, CO, under construction. The Dolores River flows below at the base of the sandstone cliffs. The flume, now defunct, was used for hydraulic gold mining. Image source: Denver Public Library call number X-60201.

FLUMES

There are many reasons to move water from one place to another, and this motivates people to get creative. A straightforward solution is to use a *flume*, which is a human-made channel for carrying water and is generally constructed of wood, concrete, or steel. In the West, people use these structures to deliver water for agriculture, mining, and hydroelectricity, or for transporting commodities such as logs. In many places, flumes simply convey water over low spots in the land.

FLUORIDATION

Most municipal water suppliers routinely practice *fluoridation*, the addition of the element fluorine to drinking water to help prevent tooth decay. Scientists and dentists hold a broad consensus that fluoridated water reduces tooth decay

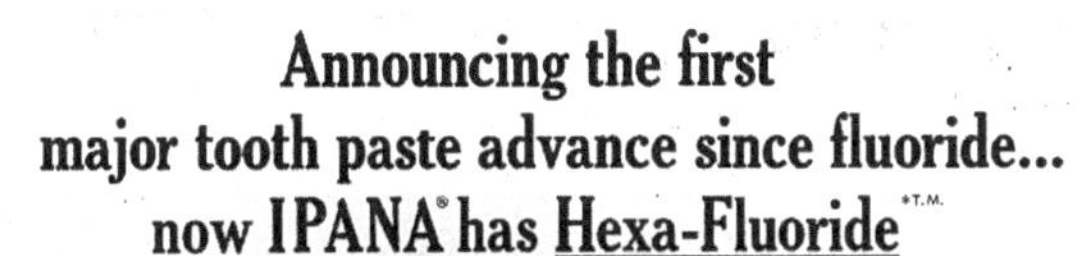

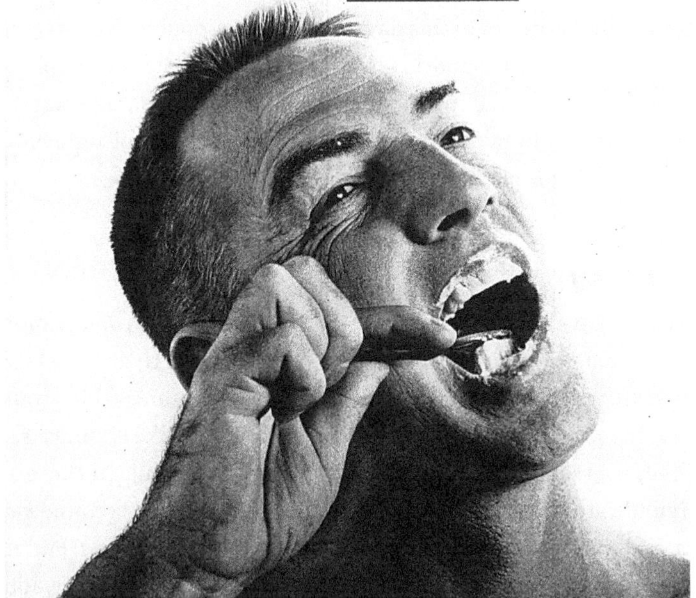

Figure 2.87. When medical authorities began adding fluoride to water, companies quickly began producing toothpaste to capitalize on the element's dental benefits. Source: *Life* (September 22, 1961): 1.

without causing harmful side effects. Nevertheless, fluoridation has had a long, storied history of controversy in the popular culture that, in many ways, anticipated the antivaccination movement of more recent years. Because of misinformation and outright lies, a large cadre of skeptics in the general public views the practice with suspicion.

When dentist Fredrick McKay moved to Colorado Springs, Colorado, in 1901, he noticed an ugly brown mottling of children's teeth in the community. McKay and another doctor, Green Vardiman Black, began studying the mottling. After several years of careful research, they deduced two interrelated facts. First, that the mottling was due to the presence of the element fluoride in the water supply, and second, that children drinking fluoridated water experienced a significantly lower incidence of tooth decay than other children. When they added only trace amounts of fluoride to water, no mottling would occur, but the teeth would remain fortified. US Army doctors eventually began studying fluoride and concluded that small amounts of fluoride would strengthen soldier's teeth, thereby improving the overall health of the fighting force. Widespread additions of low-concentration fluoride to drinking water in the post–World War II era led to dramatic decreases in tooth decay in the general public. Adding fluoride to the water supply continues to this day. Nevertheless, doubts among some people persist a half century after fluoride introduction. In 2013, for example, activists in Portland, Oregon, forced the city to drop plans to fluoridate its water, leaving Portland one of the few major American cities that still do not supplement their water supply with this element.[141]

FOSSIL WATER

Some aquifers have meager recharge rates, so low that when you pump groundwater from them, the groundwater table declines rapidly. Hydrologists often use the term *fossil water* to describe water from these nonrenewable groundwater resources. The analogy to fossil fuels is unmistakable: like oil, once an oilfield is depleted, users must turn to new sources or do without. In these aquifers, when groundwater pumping far exceeds natural recharge, the groundwater table declines, and the economic life of the aquifer eventually collapses. Furthermore, various studies estimating the age of water in slowly recharging aquifers in sites such as the Great Basin or the Denver Basin indicate that the water being pumped out originally made its way into those aquifers as far back as the Pleistocene. So indeed, this water could be accurately termed fossil water.[142]

Figure 2.88. Communities such as this one near Castle Rock, CO, rely almost entirely on fossil water from underlying bedrock aquifers. Here in the Denver Basin, as in many places across the West, communities pump their local aquifers at rates many times faster than the natural recharge rate. Even with aggressive conservation measures, it is just a matter of time before the aquifer on which these communities depend diminishes to such an extent that residents must drill additional wells elsewhere to forestall the day their original wells run dry. That, or they must find and fund the development of expensive surface water supplies to replace the wells. Photo courtesy of the author.

FRACKING

People use water in a multitude of industrial practices across the globe. So, singling out any particular industrial use may seem a bit unfair. However, hydraulic fracturing, or *fracking* as it is popularly known, of oil- and gas-bearing formations to increase production, merits special attention due to vast quantities of water consumed. In fracking, oil companies pump a mixture of

Figure 2.89. This mural about fracking located in Las Vegas, NM, integrates many of the key issues swirling around the practice: groundwater pollution, industrial economics, fossil fuel development, local and state politics, renewable energy, agriculture, and landscape preservation. Photo courtesy of the author.

chemicals and water under high pressure into wellbores to create small cracks, or fractures, in underground geological formations to allow oil and natural gas to flow into the wellbore and thereby increase production. Fracking fluids typically contain approximately 90 percent water, 9.5 percent sand, and 0.5 percent added chemicals.

The amounts of water used to frack a well vary greatly and often range from 2 to 8 million gallons (6.1 to 24.6 acre-feet). To put these quantities into perspective, if a typical western household consumes one fourth acre-foot per year, fracking a single well consumes the equivalent amount of water used by anywhere from 24.4 to 98.4 families in a given year. In a study of water use in fracking, Duke University researchers estimated that, on average, from 2005 to 2014, the United States annual water use rates were 116 billion liters per year for shale gas (approximately 188,085 acre-feet) and 66 billion liters (approximately 53,500 acre feet) per year for unconventional oil. Using the quarter acre-foot metric, that is enough water to supply the annual household needs of an American city containing 966,340 homes.[143]

Botched fracking jobs can contaminate groundwater supplies. However, linking contamination from fracking fluids to specific polluted groundwater wells is difficult to prove. Perhaps the greatest threat to groundwater is from improperly completed oil and gas wells. In poorly completed wells, fracking fluids and oil and gas products can migrate along the wellbore from the petroleum bearing formations and into shallow groundwater aquifers. If this occurs, water used for domestic and farming purposes may get contaminated. It is therefore critical that state and federal environmental agencies monitor the completion and fracking of oil and gas wells to ensure public safety and the protection of water supplies.

Moreover, significant quantities of water flows back out of the well after fracking, called *flowback*. Plus, the wells generally generate much more water from the oil- and gas-bearing strata during production, also known as *produced water*. This flowback and produced water contains chemicals from the fracking process, oil and gas residues, and is usually brackish to highly saline. Consequently, this water requires careful treatment before releasing it into the environment.

GAGES AND WATER MEASURING

With water, as one of the West's critical commodities, technologies for quantifying how much water is flowing become very important. Consequently, engineers go to great lengths to acquire water measurements on streams, lakes, reservoirs,

Figure 2.90. In this image of the Tuolumne River below Hetch Hetchy Dam near Yosemite, a hydrologist in a cable car collects data where it is otherwise too deep to wade. Shown is the masonry well, house car, cable, and stay wire of the gage. Within the stone gage house in the background, charting devices record flows and other information. Hydrologist C. J. Emerson is in the car. Image courtesy of the US Geological Survey. Public domain image.

ditches, or other bodies of water. Many federal, state, and local agencies measure streamflow for a variety of purposes such as water administration, environmental monitoring, flood preparedness, and recreation. Hydrologists calculating streamflow measure the depth and velocity of water across stream channels where they believe channel conditions remain relatively stable. They then calibrate the flow measurements and draft charts of flow versus stream height. These charts allow them to make rapid visual estimates of flow.[144]

Gages are particularly crucial in the West, where water rights determine how much water users get.[145] Initially ad hoc gaging took place around the West. Then, California began continuous stream gaging in 1878, Colorado in 1881, and the US Geological Survey and other states followed suit.[146] Records collected at gages help ensure that users take only the water to which they are entitled. Technicians often connect modern gages to the internet, allowing users to see, in real time, river flows from the comfort of their home or office.

GEYSER

Perhaps the most famous geyser in the world is Yellowstone National Park's Old Faithful. But the term itself comes from the village in Iceland that lends its name to this unique hydrologic feature. Pressured by superheated groundwater, geysers eject their water into the air when the subterranean pressure of hot water exceeds the weight of colder water above it.

In the case of Yellowstone, a shallow magma chamber heats groundwater that sources the park's famous thermal springs. Before becoming America's first national park, the thermal properties were well known and revered by the Indigenous people who lived in the region. As Americans intruded into the area in the 1860s and 1870s, rumors of the springs captured America's imagination.

Figure 2.91. William Henry Jackson made this famous photograph of Old Faithful erupting in 1872. It became the first widely distributed image of Old Faithful that the American public saw. Although the Army photographer Thomas Hine made images of Old Faithful in 1871, these went missing for more than a century. Due to the long exposure needed to make the image with Jackson's equipment, he later added white streaks to the plume by hand to enhance the drama of the photo. Image courtesy of the US Geological Survey. Public domain image.

Various artists, including the photographer William Henry Jackson of the USGS Hayden Expedition, began producing images that further fueled interest in the area. In 1872, Yellowstone was made a national park to help, among other things, protect its unique hot springs and geysers.

GILBERT, GROVE KARL

One the most influential American geologists, but little known outside academic circles, Grove Karl Gilbert (1843–1918) made outstanding contributions to hydrology, glaciology, geomorphology, tectonics, geological field methods, earthquake studies, and other branches of the earth sciences. Staff at the US Geological Survey considered Gilbert first among peers: "Gilbert was probably unsurpassed by any geologist of his time in sheer balance of mental powers."[147] Gilbert's most significant contribution was his study of hydraulic mining and its impacts on rivers of California's gold districts. His work demonstrated the effects of placer mining and showed how sediment dumped into stream channels increased the intensity of floods downstream from the mines clear to the ocean.

Figure 2.92. Grove Karl Gilbert (1843–1918) performed many seminal studies in geology, hydrology, groundwater, glaciology, and geomorphology. Image courtesy of the US Geological Survey. Public domain image.

Long before that, however, Gilbert made his name working with two of the great government-funded exploration surveys of the American West. Beginning in 1871, he served as the Wheeler Survey geologist and moved over to John Wesley Powell's expedition in 1874, surveying the Rocky Mountain region. Powell was so impressed by Gilbert's capabilities that he hired Gilbert as a geologist for the US Geological Survey in 1879, and Gilbert eventually became its first Chief Geologist (1889–1892).

During his time with the US Geological Survey, Gilbert authored the first detailed study of Utah's Henry Mountains and identified Lake Bonneville as a remnant of a Pleistocene glacial lake. Gilbert's greatest failure was his incorrect assertion that Meteor Crater in Arizona was volcanic in origin. Gilbert also performed studies of the groundwater of the Arkansas Valley of eastern Colorado and worked on glaciers in Alaska. And, in the aftermath of the great San Francisco earthquake and fire of April 18, 1906, Gilbert helped describe the earthquake and its effects on the structures of the city.[148]

GLACIER

Around 1820, two Swiss engineers, Ignatz Venetz and Johann de Carpentier, began looking at their nation's U-shaped valleys, large erratic boulders, and mixed sediments that exist far from modern glaciers. They concluded that these landforms and sediments might too result from glacial activity. They had trouble convincing anyone else of their hypothesis until they took a young German biologist and geologist, Louis Rodolphe Agassiz (1807–1873), into the field and showed him their evidence. Agassiz soon became such an enthusiastic convert to their ideas that he developed into the de facto founder of modern glaciology. For this reason, we now understand, as the National Snow and Ice Data Center puts it, that glaciers "are made up of fallen snow that, over many years, compresses into large, thickened ice masses. Glaciers form when snow remains in one location long enough to transform into ice. What makes glaciers unique is their ability to flow. Due to sheer mass, glaciers move like very slow rivers."[149] Through this capacity to flow, glaciers carve out their distinctive valleys and transport vast amounts of sediment that accumulates as moraine fields.

Agassiz described his theories in an 1840 monograph that brought him international recognition.[150] He then studied Scotland's glacial terrain. On the strengths of these accomplishments, Agassiz began lecturing at the Lowell Institute in Boston and finally as a Harvard professor. While in the United States, Agassiz assembled evidence that glaciers had modified the North

Figure 2.93. This series of images documents the retreat of the South Cascade Glacier, WA, during the twentieth century and the beginning of the twenty first. The images clearly show the effects of climate change through the rapid retreat of the ice and the formation of a glacial lake. In the photos, starting from the left with an image from 1928, the South Cascade Glacier was a majestic sight. The glacier had retreated noticeably by 1965. And by 2006, the glacier was drastically smaller. Photographs courtesy of the US Geological Survey. Public domain images.

American Continent. He wrote about the U-shaped valleys carved by ice; large glacial erratic boulders carried long distances by ice; the ice origins of parallel scratches on bedrock; and the glacial sediment pushed up as moraines. Agassiz was the first to propose that the Earth experienced an ice age during the period we now call the Pleistocene (2,580,000 to 11,700 years ago). His theories formed the basis for our recognition and mapping of the continental ice sheets that once covered North America, the Rocky Mountains and Appalachians' glacial features, and the Great Lakes' glacial origin. Had Agassiz stuck with glaciers, his work would have ensured an exalted place in scientific history. Regrettably, Agassiz stubbornly opposed the ideas of Darwinian evolution against all evidence and believed firmly in creationism. Moreover, Agassiz was staunchly racist, a fact that has so tarnished his reputation that it is nearly impossible to disengage his contributions to glaciology from his extreme scientific and moral failures elsewhere.[151] Still, Agassiz's foundational work in glaciology enabled later scientists to begin unraveling the glacial history of western North America. Grove Karl Gilbert, for example, went on to identify the glacial origins of Lake Bonneville. Harlen Bretz deduced the glacially related origin of Washington's scablands. Others mapped glacial terrain in the Rockies, Uintas, Cascades, and other ranges. Moreover, river terraces, many formed during cycles of Pleistocene glaciation, are the sites of intensive irrigation across the West.

History of glaciology aside, if we continue on our present course, the time to see a glacier in the American West may soon pass. We now know that human-induced climate change is real and accelerating.[152] Unless citizens vigorously push all of our local, state, and federal governments to take aggressive action to reduce carbon inputs into the atmosphere, the last Western glaciers will likely disappear before the end of the century, if not much sooner.

GLEN CANYON, GLEN CANYON DAM, AND LAKE POWELL

> "Glen Canyon died in 1963 and I was partly responsible for its needless death. So were you. Neither you or I, nor anyone else, knew it well enough to insist that at all costs it should endure."—David Brower[153]

As the Bureau of Reclamation puts it, "Glen Canyon Dam, rising 710 feet above bedrock within the steep, rust-colored sand-stone walls of Glen Canyon, was constructed to harness the power of the Colorado River in order to provide for the water and power needs of millions of people in the West."[154] Glen Canyon's concrete-arch dam stands at 726 feet and is the fourth-highest dam in America. Its reservoir, Lake Powell, can store over 26 million acre-feet of water. The dam has eight hydroelectric generators with a total capacity of 1,320 megawatts and produces around five billion kilowatt-hours of hydroelectric power annually. These statistics both mask and help explain the enormous political and environmental controversies that surround this desert edifice to modernity.

The US Bureau of Reclamation constructed Glen Canyon in the aftermath of a political settlement that saw the Bureau drop its plans to build a reservoir in Echo Park on the Green River in exchange for the environmental community not opposing the project.[155] Primary justifications for Glen Canyon included its role as a storage vessel for apportioning the water from the Colorado River Project, its hydroelectric potential, and its lake that would supply motorized recreation.

Figure 2.94. Glen Canyon Dam in May 2016. Photo courtesy of the author.

Some of its many environmental impacts include the loss of Glen Canyon and its scenic and archaeological resources. Other effects include enormous evaporation of water from Lake Powell, destruction of fish and wildlife habitat, its facilitation of urban development, and the modification of stream flows in the Colorado River. Glen Canyon Dam remains a lightning rod for environmentalists who call for its removal and the restoration of the canyon that the dam inundated.[156]

Climate and hydrological modeling in the Colorado River basin suggest that over the next century, water deliveries to all users plus water lost to reservoir evaporation will exceed reservoir inflows (due to reduced rainfall and snowmelt) into Lake Powell. These climate changes will lead to substantial degradation of Colorado River water management options. Current water demands on the Colorado River are already approaching the mean inflows to Glen Canyon. If the inflows decrease slightly due to climate change or if demands increase further, we can expect substantial declines in water system performance.[157] No reservoir system, such as Lake Powell, can, over the long term, meet water demands that exceed the mean inflows.[158]

GRAND CANYON DAMS

Among the most famous dams in the West are two that were never built. Proposed by Stewart Udall in the mid-1960s (recycling ideas from as early as the 1920s) as part of the Pacific Southwest Water Plan, the Bridge and Marble Dams would have flooded sections of Arizona's Grand Canyon. The proposals for these dams galvanized the environmental community in a fight to save the Grand Canyon. Engineers proposed the Bridge and Marble Canyon Dams to produce power for the Central Arizona Project. Although the US Bureau of Reclamation claimed the reservoirs created behind the dams would enhance recreational opportunities in the canyon, motorized boating was not what many Americans wanted to see within one of America's most famous national parks. Concern was so great that the Sierra Club took out full-page advertisements against the dams in the *Washington Post*, the *New York Times*, the *San Francisco Chronicle*, and the *Los Angeles Times*. Because these dams were proposed for each end of the Grand Canyon, some people derided the projects as a "dam sandwich." US Bureau of Reclamation Commissioner Floyd Dominy always denied that lobbying by the environmental community led to the scuttling of the Grand Canyon dams. Instead, Dominy blamed Admiral Hyman Rickover and J. Robert Oppenheimer for their promises regarding atomic energy. Still, the campaigns against the dams were so vigorous that the entire Central Arizona Project legislation

Figure 2.95. This is a conceptual drawing of the proposed but never built Bridge Canyon Dam in the Grand Canyon prepared by the US Bureau of Reclamation in 1949. Image courtesy of the US Bureau of Reclamation, public domain image.

became jeopardized. By 1967, the US Bureau of Reclamation gave up its proposals for the dams when the Central Arizona Project legislation was revised to include authorization for a coal-fired power plant at Page, Arizona (the Navajo Generating Plant, itself now decommissioned), to run the pumps needed within the Central Arizona Project.[159]

GRAND COULEE DAM

One of the greatest American engineering achievements, construction of the Grand Coulee Dam, began in the depths of the Great Depression and continued until its completion about six months after United States entered World War II. Nearly every aspect of this Bureau of Reclamation behemoth is immense. At 550 feet high, the 5,223-foot-long gravity dam holds back the flow of the mighty Columbia River. It impounds over 9.5 million acre-feet of water, and its reservoir, Franklin Delano Roosevelt Lake, has a surface area exceeding 125 square miles. The bureau constructed the Grand Coulee Dam to produce hydroelectric power

Figure 2.96. A composite panoramic view of Grand Coulee Dam in 2016. Photo courtesy of the author.

and to supply irrigation water to the Columbia Basin Project. Its associated irrigation works include some 330 miles of lined canals, 4 dams and thousands of miles of laterals and drains. Grand Coulee's power plant contains 33 turbines that can produce 6800 megawatts of electricity, a generating capacity that is more than three times that of Hoover Dam. Once operational, its power transformed the energy sector in the American Northwest. Its completion enabled the United States to produce the aluminum used in airplanes that were essential for the war effort. Also, the military used its electricity to enrich the uranium that would support America's production of the first atomic bomb.

Construction of the Grand Coulee Dam did not come without enormous social and environmental costs. The pool displaced many thousands of Indigenous people from their historical lands: the Spokane Reservation lost 3,000 acres, and the Colville Reservation lost 18,000 acres. More egregiously, their livelihood was severely interrupted, as the dam blocked the migration of anadromous fish such as the salmon and steelhead trout on which their culture and identity depended. In the name of conservation, Grand Coulee and other hydropower facilities on the Columbia River and its tributaries caused the fisheries of the Columbia Watershed to fall into a steep decline from which they have never recovered.[160]

GREAT AMERICAN DESERT

In 1819 and 1820, Major Stephen Harriman Long of the US Topographical Engineers mounted a scientific exploring expedition to the watersheds of the Mississippi and Missouri Rivers. During the journey, the scarcity of water they encountered deeply impressed Long. On returning, Long prominently displayed the words "GREAT DESERT" across his map of Colorado's high plains

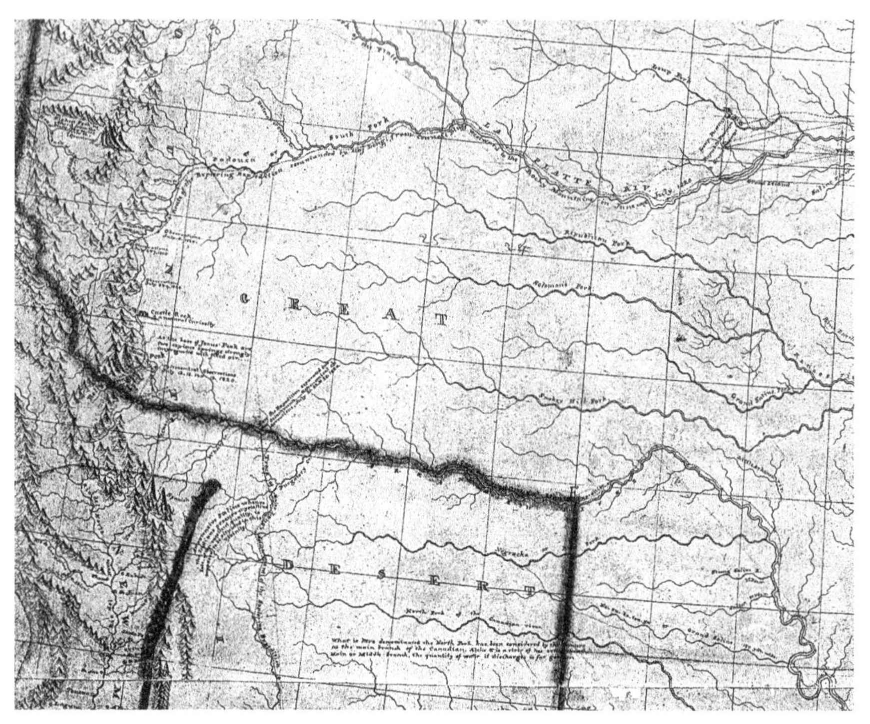

Figure 2.97. Detail of the map resulting from Colonel Stephen Long's expedition to the West in 1820–1821. On it, Colonel Long emblazoned the words "GREAT DESERT," and in doing so, set in motion the myth of the Great American Desert. Image courtesy of the Library of Congress.

between the South Platte and Arkansas Rivers. "I do not hesitate in giving the opinion, that it is almost wholly unfit for cultivation, and of course uninhabitable by a people depending upon agriculture for their subsistence," Long wrote. "Although tracts of fertile land considerably extensive are occasionally to be met with," he added, "the scarcity of wood and water, almost uniformly prevalent, will prove an insuperable obstacle in the way of settling the country." It was through these words that the notion of a Great American Desert spanning the West was born.[161]

At the time of Long's journey, Americans had not yet constructed the ditches and reservoirs that reversed the notion that the region was a vast arid wasteland. Thus, the water Long encountered was confined mainly to the well-spaced creeks that his expedition crossed along the way. Even though much of the area is technically semiarid based on modern definitions, to a traveler on horseback, the designation "desert" was likely entirely appropriate.

GREAT BASIN

During his explorations in 1842, John C. Fremont coined the name *Great Basin* when he came to understand the general configuration of the vast region in the American West that had no rivers draining to the sea. At over 200,000 square miles, the Great Basin is a hydrographic province that drains entirely within the North American continent. Within this region, all precipitation evaporates, sinks into the ground, or flows into lakes. That being so, many valley bottoms in the area contain lakes that are naturally saline or salt pans such as the Bonneville Salt Flats. Alluvial sediments within the Great Basin valleys contain vast quantities of groundwater. The Wasatch Mountains bound the Great Basin on its eastern flank, the Sierra Nevada Range on the west, the Snake River Plain to the north, and a series of lesser ranges mark the rather indistinct southern boundary. The Great Basin includes most of Nevada, about half of Utah, and parts of Idaho, Oregon, California, and Wyoming.

The term *Great Basin* is itself misleading. As part of the larger Basin and Range Province, the region consists of many small basins bounded by mountain ranges.

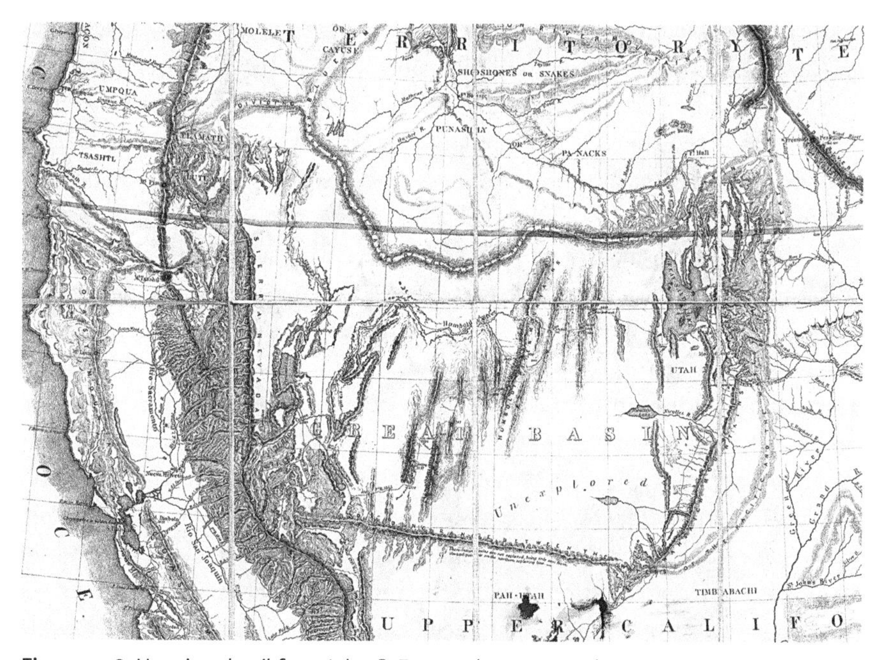

Figure 2.98. Here is a detail from John C. Fremont's 1848 map depicting the Great Basin as it was initially delineated. Subsequent mapping has drastically refined the outlines of this physiographic province. Source: *Map of Oregon and upper California from the surveys of John Charles Frémont and other authorities*, 1848. Published by the US Senate. Library of Congress, http://hdl.loc.gov/loc.gmd/g4210.ct000910 Public domain image.

Some of these small basins include those of the Great Salt Lake, Pyramid Lake, and the Humboldt Sink.

GREYWATER AND RECLAIMED WATER

Except for water from toilets or urinals, greywater is the wastewater produced from indoor activities. To utilize greywater supplies, builders must install a secondary collection system to intercept water from rooftops, sinks, showers, dishwashers, clothes washers, and other utilities that discharge water. Greywater, also known as sullage, contains fewer pathogens and little fecal matter, making it safer to handle, treat, and then reuse. Greywater systems might supply water to toilets or irrigate gardens or wetlands. As the name implies, greywater is not fit to drink but is fit for uses that do not involve direct body contact. The use of greywater makes particular sense in areas where water supplies are limited or when communities use nonrenewable water supplies.

Greywater systems have been around for many years. People started using these systems as components of "earth ships," the self-contained energy and water-efficient countercultural homes that sprouted up in the high desert areas of New Mexico (particularly near Taos) and other states. However, the adoption of greywater systems has been slow across the West. More recently, these systems have gone mainstream, and developers seeking to construct a LEED-certified building may incorporate these systems to reduce energy and water consumption footprint.

Reclaimed water, another type of water reuse, is a big step up in scale from grey water systems. In reclaimed water systems, utilities collect water that has already passed through wastewater treatment plants and redivert it for reuse rather than discharge back into a stream, lake, or ocean. This supply includes water from domestic or municipal wastewater systems, processed industrial water, and sometimes stormwater. Water users occasionally refer to reclaimed water as recycled water or reuse water. What sets reclaimed water apart from the water that utilities discharge from a wastewater processing plant is additional—often extensive—processing. Depending on the intended use, the reclaimed water undergoes other purification steps, including sophisticated filtration, reverse-osmosis processes, and high levels of disinfection to ensure that the water meets stringent water quality requirements.

A variety of issues trail reclaimed water. Derided by some as "toilets to taps," water utilities have a high threshold to cross before the general public gets over the yuck factor in using this source as part of its potable water supply. To address

Figure 2.99. Here, greywater irrigates plants in a greenhouse attached to an earth ship near Taos, New Mexico. Photo courtesy of the author.

this, some communities only use reclaimed water for non-drinking purposes, such as irrigation, industrial processes, or flushing toilets. Others use reclaimed water to recharge aquifers so that there is some physical and psychological distance between the point of discharge and the point of reuse. Some utilities have waged advertising campaigns that seek to relabel this source as a "flush to flowers" supply. Another issue affecting water reuse is the claim from downstream water right holders that when cities implement reclaimed water projects, the amount of water going downstream declines from historical levels. It's a fair critique. Some municipalities have responded by accounting for how much water moves through their systems and only use water for reclaimed purposes from transbasin sources (so they do not reduce native basin supplies). Elsewhere utilities use wastewater that would otherwise drain into the ocean and not be available for anyone else. These issues aside, the use of reclaimed water will likely increase with time as other sources become constrained or less reliable.[162]

GROUNDWATER

Subterranean water—groundwater—circulates and percolates beneath the earth's surface through the pores of bedrock, between the grains of unconsolidated

Figure 2.100. In what would be a modern-day hydrologist's dream, five artesian wells spew groundwater into the head of a canal near San Bernardino, California. (Plate 37 in Irving, W., 1900. Duty of water under Gage Canal, Riverside, CA. In Mead, Elwood, and Johnson, C.T., Use of Water in Irrigation: Report of Investigations made in 1899. Bulletin 86, US Department of Agriculture, Office of Experiment Stations, US Printing Office, Washington.)

sediment, and through fractures in igneous and metamorphic rocks. In some cases, actual underground streams flow, such as in karst terrain where limestone has dissolved away to form channels. Mostly, however, groundwater is found in porous rocks and sediments. Technically, all underground water is groundwater, but from a practical economic perspective, we mainly concern ourselves with water that can actually be retrieved for use by people. It is for this reason we mostly concern ourselves with groundwater derived from springs and wells.

In the West, many of our cities and agricultural districts rely on groundwater supplies for their persistence. Vast aquifers supply major metropolitan areas: Phoenix, Tucson, suburban Denver, many California coastal cities, to name a few. Similarly, groundwater provides the agricultural needs across considerable portions of the Great Plains via the Ogallala Aquifer, montane valleys such as Colorado's San Luis Valley, and along numerous waterways where installing pumps is more expedient than building ditch systems.

At some point in the hydrologic cycle, water finds itself in the atmosphere where it falls to the earth as rain or snow. From there, water descends directly into the ground, or first runs off into streams and lakes where it then passes into the ground. One notable feature of groundwater is that it flows rather slowly—sometimes orders of magnitude slower—compared to surface water (groundwater in karst terrain is the exception because it flows in underground channels). The water flows slowly because of the action of friction as it passes through the pores in rock or sediment. To illustrate this, hydrologists typically measure the velocity of rivers in feet per second, whereas they record the velocity of groundwater in feet per day.[163]

HAYDEN, CARL

The twentieth century was a fertile era for western politicians supporting water developments, and few were more productive than Arizonan Carl Hayden. In the early 1900s Hayden became active in the Democratic Party holding positions including County Treasurer and Sheriff of Maricopa County. Upon Arizona's ascension to statehood in 1912, Hayden won a seat in the US Congress and journeyed to Washington, where he would remain for the rest of his long career. In 1927 Hayden was elected to the Senate. From his perch in Washington, Hayden advocated for the Salt River Project and the creation of Grand Canyon National

Figure 2.101. Carl Hayden, in about 1916. Source: courtesy Wikimedia Commons. Public domain image.

Park. He supported reclamation along the Gila River. Although he initially opposed the Boulder Canyon Project that would create Lake Mead, he eventually relented when he obtained favorable amendments benefiting Arizona in the final bill authorizing the project. After that, Hayden became a booster for various New Deal water projects. During this time, he supported California's Central Valley Project, and he shepherded the authorization legislation for the Grand Coulee Dam through Congress. Eventually, he worked on legislation to bring water from the Colorado River to Phoenix and Tucson that would become the Central Arizona Project. As a capstone water development achievement, Hayden cosponsored the Colorado River Storage Act of 1956 that led to the authorization and construction of the Glen Canyon Dam.[164]

HEADGATE

The headgate is the essential starting point for ditches, acequias, and aqueducts. It is the structure we open or close to take water from a river. Headgates represent a legal point demarcating where the management of water passes from the hands of the state on rivers to private persons, cities, and industries along ditches and canals. Users may open or close a headgate to take water, but headgates rarely form a barrier for the organisms living on either side of it. Fish, algae, aquatic plants and animals, and seeds all flow freely through headgates. If these species happen to be invasive, headgates do little to stop their spread. Likewise, beneficial species, such as cottonwood trees, wetland plants, and their kin, spread down ditches and provide some ecosystem services along the way. For all of these reasons, to me, the lowly headgate is one of the essential pieces of hardware that make the western experience unique.

I find the term *headgate* a metaphor for the control of nature. On one side of the headgate, a river's water seemingly runs free and clear; on the other, it is trained to flow in ditches to serve farms, ranches, industries, and cities. It simultaneously unites or separates nature and society; indeed, the agency of humans and nature become intertwined at the headgate. Of course, if you look closely, you'll find that people carefully manage rivers, or that wild nature survives and sometimes thrives after the water passes through a headgate into a ditch. So, to be sure, it is an imperfect metaphor, but nonetheless useful for thinking about (and often debunking) the dualities that people use to set themselves apart from nature.

Figure 2.102. Headgate installations, like this on the Cordillera Ditch located on Costilla Creek in northern New Mexico, generally have similar configurations. To operate the ditch, the ditch rider first lowers the metal gate at the center. With this left gate lowered, water in Costilla Creek backs up until it begins overflowing the concrete diversion dam at the far left. Then, the ditch rider moves over to the smaller metal gate at the right to raise it and let water into the ditch. This smaller gate is the actual ditch headgate. By raising or lowering the ditch headgate, the ditch rider regulates water flow into the ditch. When I took this photo in early November 2020, the ditch rider had shut the ditch for the winter. Photo courtesy of the author.

HEADWATERS

When we speak of *headwaters*, what comes to mind is the source or uppermost reaches of a stream. Think of the Kawuneeche Valley where the Colorado River originates, or Green River Lakes out of which the Green River flows. Of course, there is a myriad of small streams and rivers that each have unique headwaters. And let's not forget that there are headwaters to the watersheds of human-made reservoirs as well.

Another way to think about headwaters is to include the various small streams that come together to form a major river. A good example is the Missouri River. It has three major stream systems that combine to create this great river. These headwater streams include the Firehole River that, in turn, joins the Gibbon to form the Madison River; Bowers Spring that joins Hell Roaring Creek, the Red Rock River, the Beaverhead River, and eventually the Big Hole River; and the

Figure 2.103. View of the Colorado River from a hill in Middle Park, Grand County, Colorado, taken in the 1920s by Louis Charles McClure. Geographers generally consider Middle Park as part of the headwaters region of the Colorado River. Image courtesy of the Denver Public Library, Special Collections. Call No. MCC-3091.

Gallatin River in Yellowstone Park. Once all these rivers connect, water in the Missouri River begins its long journey to its confluence with the Mississippi.

HETCH HETCHY

John Muir, the founder of the Sierra Club, began his appeal to the club's members to protect California's Hetch Hetchy Valley this way:

> It is impossible to overestimate the value of wild mountains and mountain temples as places for people to grow in, recreation grounds for soul and body. They are the greatest of our natural resources, God's best gifts, but none, however high and holy, is beyond reach of the spoiler. In these ravaging money-mad days monopolizing San Francisco capitalists are now doing their best to destroy the Yosemite Park, the most wonderful of all our great mountain national parks. Beginning on the Tuolumne side, they are trying with a lot of sinful ingenuity to get the Government's permission to dam and destroy the Hetch Hetchy Valley for a reservoir, simply that comparatively private gain may be made out of universal public loss, while of course the Sierra Club is doing all it can to save the valley. The Honorable

Figure 2.104. Hetch Hetchy Valley before work began on the O'Shaughnessy Dam. Image courtesy of the US National Park Service. Public domain image (Cat. #12,289)

> Secretary of the Interior has not yet announced his decision in the case, but in all that has come and gone nothing discouraging is yet in sight on our side of the fight.[165]

Of course, Muir's appeal to the Secretary of Interior Ethan Hitchcock was ultimately for naught, and the great dam was built, drowning Yosemite's sister valley. But before this happened, conservationists, led by Muir, waged a protracted political battle to protect the Hetch Hetchy.

The City of San Francisco was Muir's antagonist in this battle, for it had designs on the Hetch Hetchy Valley since its engineers first suggested the narrow lower gorge for a dam around 1882. San Francisco was undergoing rapid development during this time, and the city was struggling to secure ample freshwater supplies. However, in 1890, Congress created Yosemite National Park, America's first national park, and included the Hetch Hetchy Valley within its boundary. This designation seemed to thwart San Francisco's plans. But on April 18, 1906, a 7.9 magnitude earthquake struck the city and ignited a massive fire. These dual events decimated much of the town. As San Francisco reconstructed, the city administration redoubled its effort to build the reservoir.

For all the parties, the stakes could not have been higher: was the integrity of a national park paramount, or could water developers use their political might to induce Congress to undo the protections bestowed upon national treasures? For several years, San Francisco lobbied Congress and the Theodore Roosevelt administration. Muir and his allies argued for preservation. But when Woodrow Wilson assumed office in 1913, the tables turned in favor of San Francisco, and Congress passed a bill known as the Raker Act authorizing a grant for the reservoir that the president signed on December 19, 1913. Work for the dam began in 1914, actual groundbreaking occurred in 1919, and workers poured the last concrete in 1923.

This first great environmental battle of the twentieth century anticipated controversies over dam construction to come: Echo Park, Glen Canyon, and Two Forks, to name a few. It also represented a battle over the value of nature. Should nature be preserved in its primordial state, as John Muir asserted, or should it serve humans' utilitarian needs, as Theodore Roosevelt's chief of the Forest Service, Gifford Pinchot, argued? Ultimately, the City of San Francisco prevailed and constructed the O'Shaughnessy Dam in the Hetch Hetchy Valley. But in many ways, this dam became the original sin that animated all future fights over dams in the American West. To this day, no large dam can get built in the West without the specter of Hetch Hetchy hanging over it.[166]

HIGH SCALERS

During the era when the government constructed the large dams, the high scalers gained particular notoriety for their role in building those structures. They became especially famous with their exploits during the construction of Hoover Dam. None of the jobs performed during the construction of Hoover Dam was more dramatic and dangerous than that of the high scalers. Dangling from the cliff face with climbing ropes, these workers would scale off—that is, peel loose or weathered rock—from the canyon walls where other laborers would later attach the future dam abutment to the bedrock. High scalers used jackhammers to drill holes and then placed dynamite in the rock wall to clean loose rock from the bedrock's face. Their work was treacherous and hard, and they often grappled with temperatures above 100 degrees. For this risk, high scalers earned about 50 cents an hour. To gain employment as high scalers, the men had to be agile, strong, and fearless of heights. One mistake, and you might get tossed from your rope or crushed by falling rock. Because of their comfort with heights, many Indigenous men, including those from the Yaqui, Crow, and Navajo Nations, gained employment as high scalers. Workers noted that there

Figure 2.105. High scalers performed the dangerous task of removing rock from the left keyway for what would become the Glen Canyon Dam in January of 1960. A. E. Turner photographer. Photograph courtesy of the US Bureau of Reclamation. Public domain image.

were two kinds of high scalers at Hoover Dam: those who were quick or those who were dead. Because of the inherent danger in the job, the media heralded these men as heroes in the struggle to tame western rivers.

HITE MARINA

Originally constructed to serve the houseboats and other pleasure craft on Lake Powell, Hite Marina has become a casualty and poster child for climate change and drought in the Colorado River Basin. Built during the early 1960s, Hite Marina cashed in on the completion of the Glen Canyon Dam. Once centered on a long boat ramp to the lake, the marina became stranded well above the lake level by the early 2000s when reduced Colorado River inflows to Lake Powell caused the reservoir level to recede dramatically.

The namesake for the area, Lewis Cass Hite (1843–1914), gained notoriety by prospecting for gold and silver along the Colorado and San Juan Rivers. With no prospect for recovering lake levels in sight, workers relocated the marina's gas station and floating docks into the lake. Today the former marina site serves as a forlorn sentinel to a future of diminished precipitation and higher temperatures across the Colorado Plateau as human-induced climate change grinds on.

Figure 2.106. Here at the site of the Hite Marina boat ramp, the level of Lake Powell has dropped so far that the Colorado River now flows at the base of the cliffs in the far distance. Photo courtesy of the author.

HOHOKAM CANALS

The earliest irrigation and water management works in what is now the American Southwest were constructed by various Indigenous tribes. Perhaps the most famous are the canals that the Hohokam of southern Arizona built during different cycles from about 100 AD to 1450 AD, a span of more than 1,300 years. Their efforts resulted in the irrigation of tens of thousands of acres by constructing more than 300 miles of hand-dug canals in the lower Salt River Valley. The Hohokam developed an advanced culture that included cities and towns with ceremonial mounds, ball courts similar to those found in Mexico, sophisticated labor practices and organization, and exceptional pottery. In operating the canals, they built diversion structures, impounded water, and regularly maintained and repaired their facilities. Although they abandoned most of the irrigation works sometime after 1450, many were in such good condition when Euro-Americans arrived that they simply reconstructed the canals to irrigate anew the lands that would become Phoenix, Arizona, and its suburbs. The Hohokam

Figure 2.107. Archaeologist Emil Haury inspecting a Hohokam canal that he excavated at the Snaketown site south of Phoenix, Arizona, ca. 1964–1965. Photograph courtesy of the Huhugam Heritage Center, Gila River Indian Community.

were experts in farming in arid soils but also supplemented their harvests by hunting and gathering across the region.

Remarkably, some Hohokam canals were as many as 85 feet across and 20 feet deep. By steadily reducing the canal size along the way, Hohokam engineers ensured even water flow. To manage flow, they utilized many of the same technologies used in contemporary ditches, such as headgates, weirs, and laterals. They built their control structures using wood and stone. Archaeologists have marveled at the sheer size of Hohokam canals and the labor needed to construct these systems. For example, some canals required the excavation of over 28 million cubic feet of dirt, requiring an estimated excess of 25,000 person-days to build them. As such, these Hohokam irrigation systems remain among the most prominent engineering marvels from ancient North America.[167]

HOOVER DAM AND LAKE MEAD

One of the great engineering wonders of the modern world, Hoover Dam (formerly known as the Boulder Canyon Dam) impounds water for the Colorado River to form Lake Mead. Although planning for the dam began after signing the 1922 Colorado River Compact, construction took place during the Great Depression between 1931 and 1935, making it one of the signature job stimulus projects of Franklin Delano Roosevelt's New Deal. Its power plant output is enormous: it can produce about 4 billion kilowatt-hours of electricity per year, which is enough to serve 1.3 million people in California, Arizona, and Nevada. This hydropower is equivalent to the output of a modern windfarm containing over 650 turbines.

It took a gigantic human effort to build Hoover Dam. Authorizing the structure required an act of Congress that took many years to complete. Initially, many opponents lined up against the dam. Arizona's representatives objected because they believed that California would monopolize water from the Colorado River. Eastern politicians saw no benefit to their states. Power companies did not want competition from the federal government and labeled government involvement as socialism. *Los Angeles Times* publisher Harry Chandler, who owned some 830,000 acres of irrigated land in Mexico, rallied his paper against the project to protect his irrigation water. But ultimately, compromise legislation emerged, and in December 1928, Congress passed the Boulder Canyon Project Act authorizing construction of Boulder Dam, the creation of the All-American Canal, a division of water between California, Arizona, and Nevada, and $165,000,000 for construction.[168]

Figure 2.108. Hoover (Boulder Canyon) Dam and Lake Mead in April 1938. Courtesy of the US Bureau of Reclamation. National Archives and Records Administration. Public domain image.

To erect Boulder Dam, workers needed housing, and Boulder City, Nevada, a town of five thousand, emerged from the desert. They constructed roads, strung powerlines, and completed railway extensions. A consortium known as the Six Companies undertook the construction itself. Because of the Great Depression's severity, the Bureau sped up work to employ as many people as possible. A myriad of tasks was necessary: they had to prepare everything from the food supply, communications, high scaling, and concrete batch plant construction. Plus, the builders had to blast tunnels to lead water to the turbines. Not only that, they had to manufacture the turbines themselves, design the power plant and its building, and of course, construct what was then the largest concrete arch dam in the world. With all this work going on, laborers poured the first bucket of concrete in the dam itself on June 6, 1933, and 4,400,000 cubic yards of concrete later placed the final bucket of concrete in the main dam on May 29, 1935.

The resulting structure was a gravity arch dam just over 726 feet tall and 1,244 feet long. The power penstock and outlet system contain four 395-foot-tall intake towers, each controlled by two 32-foot diameter cylinder gates. Water entering these towers flowed through the abutments in a series of tunnels and sixteen

13-foot diameter power penstocks leading to a powerhouse set on each of the dam's wings. Like everything else about this vast structure, the huge spillway can pass 400,000 cfs of water. Impounded behind Hoover Dam is Lake Mead, which can contain about 32,000,000 acre-feet of water in a lake that is over 110 miles long. From a societal perspective, Hoover Dam demonstrated an exceptionally high degree of cooperation between engineers, architects, artists, hydrologists, politicians, and bureaucratic supervisors, construction contractors, and workers. This structure displayed for all the world the organizational prowess of an emerging American superpower.

HUNDREDTH MERIDIAN

Try as we may to ignore it, change it, or adapt to it, the arid western climate represents an overarching physical reality that humans must cope with if they are to survive or thrive in the West. As Wallace Stegner put it, "With local and minor exceptions, the lands beyond the 100th meridian received less than twenty inches of annual rainfall, and twenty inches was the minimum for unaided agriculture. That one simple fact was to be, and is still to be, more fecund of social and economic and institutional change in the West than all the acts of all the Presidents and Congresses from the Louisiana Purchase to the present."[169]

The 100th meridian itself is a north-south band of longitude. For a point of reference, Dodge City, Kansas, straddles the 100th meridian. But it is not the geographic demarcation that's important. Rather, for anyone interested in western water, the 100th meridian is generally coincident with, and therefore a convenient name for, the important hydrographic transition from regions to the east that have annual rainfall exceeding 20 inches to regions to the west where rainfall is consistently less than 20 inches.

Travel east of the 100th meridian and you will start encountering farmers who dig ditches to convey water away from their fields, lest their crops rot from too much moisture. Head west from the 100th meridian, and you will start finding ditches, canals, and acequias that farmers use to irrigate crops that would otherwise die for want of sufficient water. It is this climatic fact that makes civilization dependent on irrigation in the West. As Stegner noted, beyond the 100th meridian, something "more than individual initiative was needed to break the wilderness."[170] East of there, a prospective migrant might clear a field by themselves or with family help. However, in the arid West, agriculture required sophisticated organization, collective effort, and capital to build the hydraulic

Figure 2.109. When Euro-Americans traveled west beyond the 100th meridian, they encountered vast semiarid prairies. Coming from the humid East, they initially saw this region as a vast desert. As shown here, the range was particularly harsh for travelers traversing West Texas and eastern New Mexico. From day-to-day, there was no guarantee that they could find water for themselves or their stock. The Spanish called this land the Llano Estacado, or "Staked Plain," when exploring here in the 1600s and 1700s. Rural electrification, well development, and ditch-digging made many of these formerly inhospitable lands habitable. Image courtesy of the author.

systems needed for success. Although some dryland farming is possible—the dryland wheat industry comes to mind—irrigated crops have a far higher investment return than dryland farming.

Beginning with Stephen Harriman Long, nineteenth-century travelers generally labeled the lands west of the 100th meridian as "desert." This proved an unfortunate designation, as the reality is far more complex. Some areas are true deserts, but far more land is semiarid, and there are also pockets of hyperarid, subhumid, and even humid lands. Moreover, the far Northwest sees some of the highest rates of annual rainfall on the continent. However, the idea that the western United States was one vast desert became stuck in most Americans' minds and imaginations throughout the majority of the nineteenth century. When American travelers observed Mexican and Indigenous peoples in the West irrigating land to grow crops, they realized that they too could do this and successfully build towns, cities, and agricultural districts.

In 1879, John Wesley Powell published his famous *Report on the Lands of the Arid Regions of the United States* that identified the 100th meridian's critical significance to the American West. This report forever tied Powell's name with the 100th meridian. Powell sagely advised, "It is not wise to depend upon rainfall where the amount is less than 20 inches annually."[171] But American colonists already knew this and had been busy constructing the hydraulic infrastructure and legal systems to support an agricultural economy for over twenty years. Still, Powell's writings spotlighted the issues affecting the West for prospective migrants and eastern politicians alike.

HYBRID FRESHWATER ECOSYSTEMS

Numerous lakes and streams are simultaneously natural and social in origin. Although natural, many freshwater ecosystems are affected by ongoing ecologic, hydrologic, chemical, and geomorphic modifications produced by human activity. Our long history of human occupation, exploitation, and development underscores how difficult it is to distinguish between human and natural processes. People and nature were both active participants in creating these ecosystems. It may not be possible, and perhaps even detrimental, to try to delineate a bright line between the natural and the artificial when discussing many of the West's lakes and streams.[172]

A useful term for describing these sites, *hybrid freshwater ecosystems*, captures the idea of human-derived freshwater ecosystems. In these rivers and lakes, it's clear that their ecosystems cannot be understood apart from the cultural forces that created them. Human-induced changes to our streams and the outright construction of lakes make it difficult, if not impossible, to distinguish between natural systems and those that are more or less derivative of human action.

Hybrid freshwater ecosystems arise from complex interactions between people and the environment. Here, nature and culture intermingle to create hybrid geographies that are inescapably partial, provisional, and incomplete. Culture and nature are not polar opposites but rather intimately and variously linked. Human agency affects nature and vice versa. Take Lake Powell as an example. Although built by humans, it nonetheless has developed an aquatic ecosystem through the action of natural forces, something that had not previously existed in the Utah desert, but still natural in the sense that nonhuman agency accounts of much of the fauna and flora present in the lake.[173]

Figure 2.110. The wetlands at Tule Lake Sump, located within the Tule Lake National Wildlife Refuge, provides an excellent example of a hybrid freshwater ecosystem. Operated by the US Fish and Wildlife Service, managers at the refuge manipulate water deliveries to the sump to mimic a natural hydrologic regime. In doing so, they create habitat to benefit the full array of birds present, not just waterfowl. Refuge management includes irrigating approximately 17,000 acres of land for the many species that are present. B.D. Glaha, photographer. September 1941. Photograph Courtesy of the US Bureau of Reclamation. (KP-1001-R2, public domain image).

HYDRAULIC MINING

One of the great disasters to befall American rivers seems but a distant memory today. Starting with the 1849 California Gold Rush, miners quickly developed industrial-scale machines to extract gold from placer deposits. These machines sound innocuous enough, but by 1853 they had engineered technologies to wash untold tons of sand and gravel through sluice boxes to concentrate gold. They then introduced mercury into the sluices to create gold amalgams to remove the gold. The process of hydraulic mining was quite simple. Water delivered to the placer was projected at high velocity and pressure onto an excavation to create a slurry that washed away tons of boulders, gravel, dirt, and the gold it contained. Eventually, miners used robust canvas hoses fitted with iron nozzles to direct water at the placer. The largest hoses, also known as Monitors after a particular

Figure 2.111. This photo of a hydraulic mining operation in California, probably taken during the 1870s, shows the destructiveness of the process. The immense pressure of water coming from the water cannon could wash away entire hillsides of sediment in short order. https://www.sierracollege.edu/ejournals/jsnhb/v2n1/monitors.html. Public domain image.

brand, were enormously powerful. California historian Hubert Howe Bancroft reported that an eight-inch Monitor could project 185,000 cubic feet of water per hour with a velocity of 150 feet per second. These devices were so substantial that boulders would ride on the stream, and men could be killed by the water's force from 200 feet away. The quantity of water the miners used was also staggering. At California's North Bloomfield mine alone, some sixty million gallons of water per day were used. These devices led to a wholesale reworking of California's streambeds that remain modified to this day. Much of the sediment washed away by the Monitors clogged stream channels for miles downstream. The clogged streams, in turn, raised the elevation of streambeds to such an extent that devastating floods impacted downstream cities including Sacramento.[174] Sedimentation became so extreme that a "titanic struggle erupted as farmers in the (Sacramento) valley and their allies in the towns along the rivers fought to restrain the miners of the mountains." In 1884, a federal court banned hydraulic mining. That injunction stood until 1893 when Congress restored the practice.[175]

Although pioneered in California, gold seekers used hydraulic mining across the West. In many placer mining districts, vast piles of gravel remain in mute testimony to a nineteenth-century technology.

HYDROGRAPH

Hydrographs chart the flow of water over time. As such, they contain a considerable amount of information about rivers, wells, canals, and pipelines. Long-term hydrographs, charting decades or centuries of data, provide information about the regional climate, reliability of water supplies, changes in watersheds, and even climate change. Short term hydrographs, perhaps a year or less, provide information about seasonal weather patterns, spring snowmelt, water-use information, and individual storm events. Yearly maximum flows help determine the size of flood or drought events (for example, one hundred–year flood, five hundred–year drought). Baseflow data from hydrographs provide information about local groundwater conditions. The shape of the hydrograph often relates to watershed conditions such as forest cover, grasslands, or even impervious cover like pavement or bedrock. When scientists calibrate a hydrograph to gage height, they can obtain valuable information about flood or drought conditions. Likewise, when they calibrate a hydrograph to rain gage information, hydrologists may estimate how fast a massive storm might manifest as a flood.

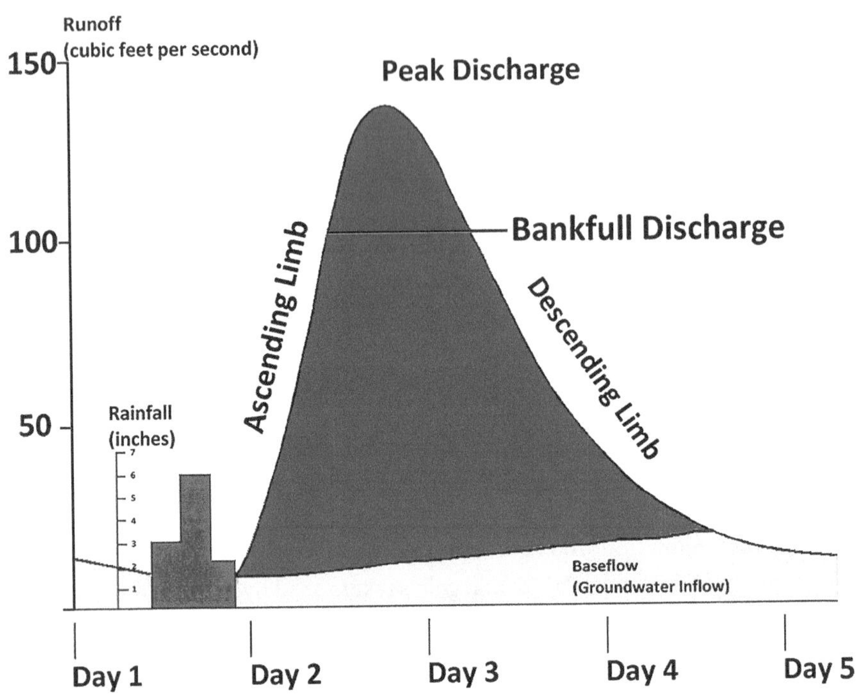

Figure 2.112. Here is an example of an idealized hydrograph showing its various components juxtaposed to information from a storm event.

HYDROPOWER

It took well over one hundred years and the cumulative work of a long list of scientific geniuses for the promise of electricity to come to fruition. Building on this foundation, in 1895 Nicola Tesla and his partner, George Westinghouse, built the world's first hydroelectric power plant at Niagara Falls. This plant and a host of other inventions led to the construction of the modern power grid. Tesla's greatest invention was to generate electricity by spinning turbines with running water. In the American West, entrepreneurs quickly implemented Tesla's technology to harness energy.

Engineers call electricity generated through the gravitational force of flowing water *hydroelectricity* or *hydropower*. In the thirteen Western states, hydropower accounts for approximately 22 percent of all electricity generated. California leads the way with over 24,000,000 megawatts (MW) produced in 2013.[176] Washington follows with large projects on the Columbia River, such as the Grand Coulee and Chief Joseph Dams. The Colorado and Columbia River

Figure 2.113. These are the hydroelectric penstocks for Chief Joseph Dam, Washington, in June 1958. In this photo, there are 16 penstocks, each of which is 52 feet in diameter. At the right is a penstock under construction. US Corps of Engineers eventually built a total of 27 of these penstocks at the dam. The dam can produce approximately 2,600 megawatts of electricity. Photograph by Lawrence D. Lindsley. Image courtesy of the University of Washington Libraries, Special Collections Negative Number Lindsley 5584.

Basins are the most heavily exploited regions for hydroelectric production in the West. Although renewable, hydropower requires the diversion or impoundment of water and consequently blocks the migration of river fishes (such as salmon and pikeminnow) that require long stretches of free-flowing rivers to survive. Hydropower tends to be quite valuable because power producers can release water on demand to produce electricity. This flexibility allows power companies to produce electricity at times of peak demand. Conversely, power companies use coal and nuclear for base load generation, as these power plants are slow to respond to rapid changes in electric demand.

INDIGENOUS WATER RIGHTS

Indigenous water rights arise from and are ostensibly protected by a 1908 Supreme Court decision that's referred to as the Winters doctrine. The Winters doctrine states that when Congress reserved land for an Indigenous reservation,

Figure 2.114. Three Akimel O'odham farmers, Oscar Walker, James Vanico, and Olinge, resting from work on irrigation ditches, ca. 1900. Image source Wikimedia Commons. Public domain image.

it also reserved water in the amounts needed to fulfill the purpose of the reservation. The Winters doctrine stems from a dispute near the Fort Belknap Reservation located in the then Montana Territory. The United States created the Fort Belknap Reservation in 1888 for the Gros Ventre (Aaniiih) and the Assiniboine (Nakoda) tribes. At the time, the government sought to transform Indigenous people from "a nomadic and uncivilized people" into "pastoral and civilized people" by providing them appropriate reservation lands. By 1905, water shortages resulted in Indigenous people filing a lawsuit (*Winters v. United States*) to enforce tribal rights to water against white water users who were diverting water. In its decision, the US Supreme Court explained that lands provided to develop agrarian societies "were arid and, without irrigation, were practically valueless." An interesting aspect of the Winters decision is that it "superimposed a judicially implied federal water right" on a state system based on prior appropriation.[177]

Unfortunately, the Winters decision did not specify a formula to identify the amount of water any particular reservation needed, which has led to seemingly unending litigation and different courts applying different standards to

Figure 2.115. Ojibwe artist Isaac Murdoch made what is perhaps the most recognizable symbol of resistance to the Dakota Access Pipeline with his use of the Thunderbird Woman legend on his poster. Image courtesy of Isaac Murdoch.

quantify tribal water rights. Additionally, the Winters doctrine has generally led to giving Indigenous people water on paper. Still, the resources needed to develop this water are often nonexistent, which has led to chronic water supply shortages on many reservations. Tribal water rights have become known as "reserved water rights" because of the necessity for water as part of creating reservations. Furthermore, since most Indigenous reservations predate virtually all of the surrounding farm, ranch, and cities that grew up around Indigenous lands, these rights often have the most senior appropriation dates on the rivers from which a tribe gets its water.

Indigenous communities continue their struggle to preserve their water rights and secure supplies of clean water. For example, take the construction of the Dakota Access Pipeline across ancestral treaty lands of the Hunkpapa Sioux in North Dakota near the Standing Rock Reservation. The Standing Rock Sioux Tribe learned in July 2016 that the US Corps of Engineers had approved the construction of the oil pipeline within one-half mile of the reservation boundary on lands containing burial and sacred sites important to the tribe. Protestors quickly erected camps at the Dakota Access Pipeline site to press for relocating the route or canceling the project altogether. The tribe is particularly concerned about the pipeline's impacts on nearby water supplies given catastrophic leaks in similar pipes in recent years. In resisting the pipeline, tribal members and their supporters have been arrested, subjected to heavy-handed police tactics including militarized water cannons to suppress protests, and other intimidating tactics intended to end opposition to the construction project. Various legal battles ensued, and led to a legal victory by the Standing Rock Sioux Tribe in 2020 that struck down the US Corps of Engineers permit as violating the National Environmental Policy Act. In February 2022, the US Supreme Court agreed, and as a result, the US Corps of Engineers must complete a comprehensive environmental review of the project.

Water remains a critical concern for Indigenous people across the West. For them, water is both a spiritual and physical necessity. Many view water bodies as living entities. Numerous tribes face grinding poverty and ongoing challenges to provide their members with reliable water supplies.

INFILTRATION AND ARTIFICIAL RECHARGE

Across the landscape, rainwater or snowmelt soaks into the ground and percolates downward through unsaturated earth. Eventually, this water reaches the groundwater body. This water joining the groundwater body is recharge water and adds to the volume of groundwater stored within the aquifer. As more groundwater accumulates, the groundwater level rises.[178] Many factors, both natural and human, affect the groundwater level. Groundwater naturally discharges into lakes, streams, or the oceans at the same time that people pump water for a wide variety of uses. Suppose you think about all of the inputs and outputs to the groundwater system. In that case, it begins to resemble an accounting sheet, and hydrologists refer to this accounting as the groundwater budget. How we manage this budget directly affects the long-term sustainability of the aquifer. Some pumping will keep the account in the black; too much pumping will force the account into the red.

Additionally, aggressive groundwater pumping invariably leads to steep declines in groundwater levels. Consequently, various technologies to increase groundwater levels have assumed critical roles in managing aquifers. Artificial recharge, the process where surface waters are put into groundwater storage, also known as aquifer recharge, is an important strategy for addressing water level problems. Water used in artificial recharge often comes from excess surface water supplies that are available during wet years or during the spring runoff when abundant water supplies are available. This allows water managers to use an aquifer like a reservoir to store water for later use, such as in a dry year or during a different season. A management technique known as *conjunctive use* is the coordination of surface and groundwater supplies through targeted artificial recharge and withdrawals to create water supply systems with higher resiliency and drought tolerance. Like any water supply system, the sustainability of aquifers using artificial recharge rests on the long-term balance between inputs and withdrawals.

Figure 2.116. This photograph shows the Tonopah Desert Recharge groundwater infiltration basins used to induce artificial recharge along the course of the Central Arizona Project supply canal. Public domain image, courtesy of the Central Arizona Project. Source: https://storymaps.arcgis.com/stories/70f626809fd84e228c0f32aef4222dda.

As we continue pumping our aquifers, groundwater supply and sustainability problems mount. This pumping has led hydrologists to consider ways to stave off the day when various aquifers become unusable. Beyond the obvious solutions of reducing pumping and managing demand, the practice of aquifer recharge offers the possibility of extending aquifers' life. As the word implies, recharge is the replenishment of an aquifer by the absorption of water. What makes this process fascinating is that it can happen in many ways. Rainwater seeping into the ground, water escaping from a stream into the ground, water flowing from fractures into the ground, and water injected into an aquifer by people are all types of recharge. If the recharge rate is higher than the rate of discharge and pumping, then an aquifer gains water. Once the discharge rate (both natural and human pumping) exceeds that of recharge, the water table will drop. Unintentional recharge from poorly plugged oil and gas wells or storage tanks at industrial sites can introduce pollutants into aquifers. The recharge rate is a critical consideration for groundwater users in determining how much water they can extract from an aquifer. As such, recharge remains an important topic for water managers across the West.

INJECTION WELL

People are always seeking creative ways to dispose of hazardous waste without spending a lot of money actually to treat it and make it nonhazardous. And what better place to put wastewater than underground, where it is out of sight and out of mind. And to get rid of liquid hazardous waste, one of the easiest ways to do this is to use an injection well, which engineers specifically design to pump (inject) wastewater directly into the ground. By disposing of wastewater in this manner, the pollutants dissipate into geologic formations far removed from human contact or ecosystems on which people rely. Engineers must carefully build the well so that the wastewater does not leak into shallow aquifers that people need or seep into surrounding streams, lakes, and wetlands. However, if the injection well is poorly constructed (which happens more than water managers care to admit), for example, through a faulty installation, waste might

Figure 2.117. In 1961, engineers completed a disposal well at the Rocky Mountain Arsenal in Adams County, CO, to inject highly contaminated industrial fluids, such as waste from this sarin nerve gas manufacturing facility, into the deep earth. The high injection pressures triggered multiple earthquakes that people felt across the Denver area. Deep disposal of waste and hydraulic fracking of oil formations remain controversial in part because of this project. Public domain image courtesy of the Library of Congress. Image call number HAER COLO,1-COMCI,1DB.

pollute shallow aquifers. If the disposal company pumps the waste at excessively high pressures or volumes, the seals in the well might fail, or the surrounding formation might fracture spontaneously, causing earthquakes. This famously occurred in the 1960s at the Rocky Mountain Arsenal, north of Denver, when high-velocity pumping of severely contaminated industrial wastewater into bedrock induced earthquakes. Pressure injection began in March 1962, and shortly after that, seismologists recorded 710 earthquakes in the Denver area. According to geologists studying these events, most of these earthquakes had epicenters within a five-mile radius of the Arsenal well. The frequency of the earthquakes seems related to the wastewater volume and pressure. Before injecting waste, the area was seismically quiet, and once pumping ended, the tremors stopped.[179]

Because of wastewater injection and the related fracking of oil- and gas-bearing formations, incidences of groundwater contamination and earthquakes have increased. This contamination has created a severe public backlash against injections, and although it continues, the controversy it generates is unlikely to abate anytime soon.

INSTREAM FLOWS

> "I think that another issue that has struck me . . . is the complete inability of our system, our Western water systems, to appreciate the value of instream uses. We still operate under this delusion that the highest and best use of water is to take it out of the river and to spread it on a field or treat it and drink it."
> —Daniel P. Beard

As the Prior Appropriation Doctrine spread around the West in the nineteenth century, it was this legal system that facilitated the systematic taking of water for development. And that development included all the beneficial uses of water as it was understood to be in the minds of nineteenth-century people. For them, beneficial uses meant removing water from rivers to grow crops, diverting water to support mining and industry, and sending water to homes and cities. Water users of the nineteenth and early twentieth centuries saw water in the environment as wasted water if someone did not actively use it. *Instream flows*, or water used for in-channel environmental benefit, would have been a completely foreign concept to the nineteenth-century water user. Nineteenth-century people went about appropriating water, getting decrees, and removing water from western streams, not keeping water in creeks for things like plants and fish.

It wasn't until the mid-twentieth century that evidence for the environmental damage due to water diversions began to mount and eventually became

Figure 2.118. Although states have established instream flow water rights for many western streams, these rights possess priorities that are usually junior to nearby agricultural rights. Moreover, some states have done little to protect instream flows. For example, New Mexico has formally granted fewer than five instream flow water rights, giving it the worst track record for safeguarding instream flows in the West. Fortunately for the lone fisherman pictured here on the Rio Grande near Arroyo Hondo, NM, water deliveries must first pass through this scenic gorge before reaching agricultural and municipal water users. Photo courtesy of the author.

impossible to ignore. By the time the environmental movement of the 1960s and 70s came into full swing, many people now understood that water for the environment itself was a critical beneficial use. Beginning in the 1970s, many state legislatures responded to river degradation by passing laws that recognized the benefit of instream flows and permitted people or state agencies to appropriate water for instream purposes. These laws helped establish minimum streamflows needed to protect fish and wildlife and allowed state agencies to manage water from various projects so that baseline amounts of water would remain in rivers. It's important to note that decrees for instream flows date from the 1970s or later, whereas those claiming agricultural, industrial, and domestic water decrees had over one hundred years' head start in getting their uses approved by the courts and administrative agencies. In other words, contemporary instream flow decrees are often far junior to all the other senior uses. Said differently, if a stream contains senior agricultural water rights, the junior instream flow rights will remain out of priority, meaning that the stream might remain dry for large portions of the year.

Accordingly, to reestablish flows in a heavily used river, agencies must often buy valuable old rights and transfer those rights to instream flow purposes. But, acquiring rights comes at great expense. Alternatively, agencies might employ administrative procedures (such as the timing of water deliveries) so that individual reaches of a river maintain an instream flow. Overall, these programs are an essential tool for improving stream conditions around the West.[180]

INTERSTATE WATER COMPACTS

When the British Crown ruled America, it often had to mediate disputes and controversies between its colonies. Colonists usually resolved disputes through negotiation that the Crown would ratify. Once the United States gained its independence, the Constitution's framers adopted this model for the new country.[181] The framers wrote, "No State shall, without the Consent of Congress, . . . enter into any Agreement or Compact with another State."[182] Many states have since used this Constitutional provision to negotiate interstate water compacts to allocate the water in a river flowing between them. In other words, an *interstate water compact* is a binding legal contract between two or more states that defines rules for water distribution between them. Many people refer to compacts as the "law of the river." A compact is the critical legal document for distributing water among states, but there are many more decrees, legal settlements, and legislative acts, that, taken together, constitute the full law of the river for each waterway. As an analogy, a treaty is a binding legal agreement between two or more countries, whereas, in a domestic context, a compact is a binding legal agreement between two or more states. For an interstate water compact to become binding, Congress and the participating states' legislatures must approve (or ratify) it. A compact is, therefore, simultaneously both federal and state law, enforceable against all water users in those states. If disagreements arise between states or the federal government regarding the compact, the aggrieved entity must file a lawsuit directly with the US Supreme Court for relief. Likewise, in the absence of a water compact, two states may sue each other, and this, too, is heard by the US Supreme Court. This kind of suit is perilous for all parties because the US Supreme Court will eventually issue a nonappealable equitable apportionment decree that has the force of federal law. These equitable apportionment decrees

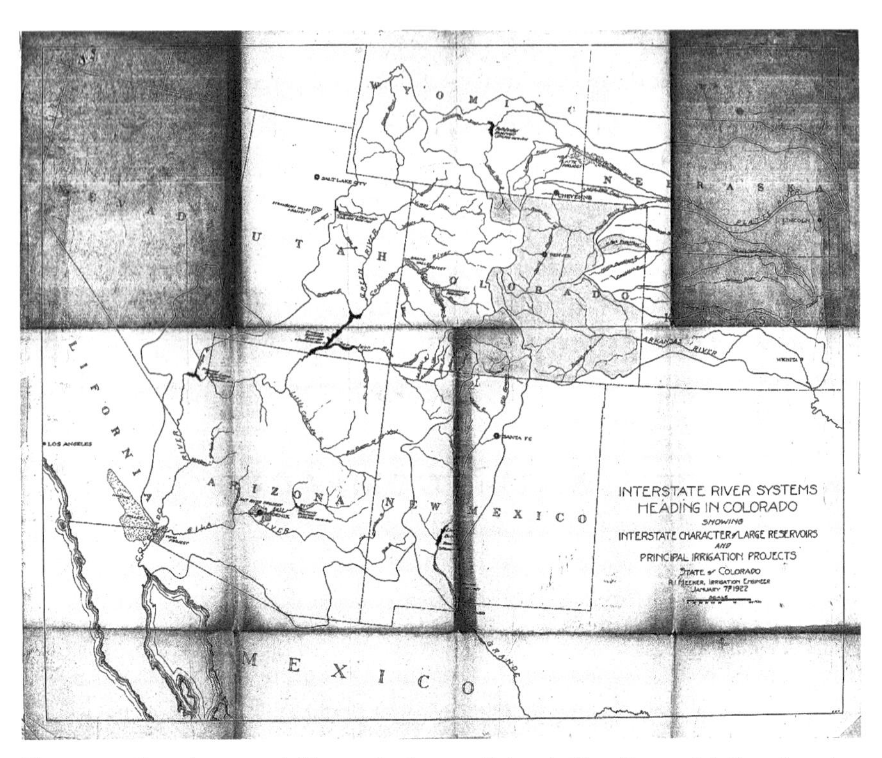

Figure 2.119. Negotiators used this map for the 1922 Colorado River Compact deliberations. Image courtesy of the Ditch Project.

remain binding on the states until such time that they sit down and negotiate a water compact that is approved by their legislatures and Congress. Of course, the winning state has little incentive to sit for further negotiations.[183]

In practice, compacts are the territory of high stakes politics, interstate conflict, and settlement, and are the decisive documents dictating whether cities and metropolises will grow or wither. The most famous of all compacts is the Colorado River Compact, concluded in 1922 at Bishop's Lodge, New Mexico. In it, the states of California, Arizona, Nevada, New Mexico, Utah, Wyoming, and Colorado agreed on the apportionment of the river. At the time of its ratification, water users believed that approximately 15,000,000 acre-feet of water flowed down the river per year. Of that, the compact allocated 7,500,000 acre-feet to the states of California, Arizona, Nevada, and another 7,500,000 acre-feet to New Mexico, Utah, Wyoming, and Colorado. In 1944, the United States and Mexico entered into an international treaty that allocated another 1,500,000 acre-feet annually to Mexico.[184] It turns out, however, that hydrologists measuring the average flow of water in the Colorado River flow did their work during twenty of the wettest years on record. Since then, it has become apparent that considerably less water is present on average, so the allocation of water between what is now known as the Lower Basin and Upper Basin states has become quite contentious.

INVASIVE SPECIES

The whole conversation about invasive species is fraught with difficult questions containing ill-defined answers relating more to values and politics than concrete scientific facts. In simplistic terms, an invasive species is a species that has been introduced from one ecosystem into another by human action. With no natural predators present, invasive species may reproduce prolifically. The result is destabilization, damage, or even collapse of the ecosystem into which it has entered. Think of kudzu in southeastern forests as a poster child for the potential damage that an introduced species might produce. In the West, damaging invasive species have overtaken many lakes and waterways. These species include Russian olives, tamarisk, zebra mussels, quagga mussels, and Eurasian watermilfoil, to

Figure 2.120. This *Stop Aquatic Hitchhikers* sign at a boat ramp on the Colorado River near Moab, Utah, is directed at the various invasive species of mussels that have spread to western waterways. The sheer number of invasive species seems to be ever-expanding. Here quagga and zebra mussels, along with New Zealand mud snails, aquatic plants including Eurasian watermilfoil, various old-world thistles, tamarisk, Russian olives, exotic diatoms, and many other species have dramatically modified river and lake ecosystems.

name a few. Once present, these species have spread across western waters and have altered the original ecosystems to such an extent that the native ecosystem might become wholly unrecognizable. Beyond these extreme examples, people have introduced a multitude of species into western waterways that might not be deemed as invasive by some but undoubtedly invasive by others. Consider the European brown trout. This trout is a species that Euro-Americans have actively introduced into waterways (even building fish hatcheries to do so) that fits the bill of invasive. However, it would not easily get pigeonholed as invasive because of its desirability as a sport fish.

Many philosophical questions revolve around the notion of native versus introduced. These questions include how long a species must be present before considering it native: one generation, two, a thousand? What about distance: must it come from another continent, region, or watershed to deem it introduced? How much displacement of endemic species is acceptable before it is considered invasive (10 percent, 20 percent, 90 percent)? Then there is the mode: if an exotic species rides in on a hurricane, does that make it somehow more natural than if it came by way of airplane?

What if the introduced species is economically beneficial but spreads widely? The brown trout is a good example. European earthworms provide a fine example, too, even though they are not an aquatic species. After all, since the earliest days of European colonization in the Americas, farmers have introduced earthworms into their soil.[185] What about an otherwise endemic riparian species that then spread out across formerly dry lands due to ditch digging? The endemic Ute ladies-tress orchid, listed on the endangered species list, has spread out onto human-created wetlands due to ditch construction. I think most of these questions can only get answered, and then only somewhat unsatisfactorily, by making arbitrary value judgments.[186]

IRRIGATION

It is hard to understate the magnitude of change in the American West attributable to the introduction of irrigated agriculture. Irrigation, which is the direct application of water to plants, is only possible with infrastructure that diverts water from rivers, lakes, and aquifers. Due to a natural soil moisture deficit and limited precipitation, people have constructed vast networks of ditches, tunnels, dams, and other structures to send water to crops that would otherwise not grow in the West's arid climate. As John Wesley Powell wrote in 1879, in much of

Figure 2.121. This is a screen grab of a satellite photograph showing part of California's Imperial Valley. The Cartesian grid of irrigated fields hints at the magnitude that irrigation development has altered western landscapes. Screen capture of image courtesy of Google Earth.

America west of the 100th meridian, rainfall is consistently less than 20 inches per year.[187] If you want to eke out an existence as a farmer west of the 100th meridian, irrigation becomes essential. In essence, to make a go at farming in the West, having water rights and the infrastructure to deliver that water is one of the fundamental requirements for success.

Because people have over-appropriated many rivers, there is often high demand for the limited water available. Farmers have become increasingly innovative in reducing water waste and delivering water to crops. Initially, farmers applied most water through direct flood irrigation. Flood irrigation is the least efficient method of providing water, but it has resulted in the creation of incidental wetland and riparian environments. In many areas, piped water and sprinkler systems have reduced water losses and get supplies to thirsty crops. But these systems often spray water and lose considerable amounts to evaporation. In areas where irrigation water is most valuable, or where few water rights are available for lease or purchase, farmers must take additional steps toward efficiency. One step is often the installation of expensive central pivot or drip

Figure 2.122. Here is a farmworker adjusting irrigation flows to cantaloupe plants in California's Imperial Valley. The photographer Dorothea Lange took this photograph on assignment in March 1937 for the US government's Farm Security Administration (FSA) program, an agency formed during the Great Depression to provide aid to impoverished farmers. Library of Congress LC-USF34-016173. Public domain image.

irrigation systems that deliver water straight to the roots of crops. The tradeoff is cost versus efficiency.

LAHONTAN CUTTHROAT TROUT

Pyramid Lake in Nevada is the home to the largest cutthroat trout in the world. Here, the giant Lahontan cutthroat trout (*Oncorhynchus clarkii henshawi*) can live up to 14 years, grow to 50 inches, and can weigh 20 pounds, with some even approaching 40 pounds. They reside in the high alkali waters of Pyramid Lake and regularly migrated 114 miles in the Truckee River between Pyramid Lake and Lake Tahoe until that was interrupted by the construction of dams. Lahontan cutthroat trout were formerly widespread throughout Pleistocene Lake Lahontan, and when the lake evaporated at the end of the Pleistocene, residual populations remained in the Truckee, Carson, Walker, Humboldt, and several other streams in California and Nevada.[188]

Figure 2.123. An angler holds a Lahontan cutthroat trout at Pyramid Lake, NV. The Lahontan cutthroat trout is one of the premier sport fishes in the West. Public domain image courtesy of the US Fish and Wildlife Service. Photo by Joanna Gilkeson. Cropped from original for clarity.

Lahontan cutthroat trout have endured a variety of threats, top among them physical fragmentation of its habitat from water projects and biological fragmentation from introduced species. Moreover, crossbreeding between Lahontan cutthroat trout and introduced trout has impacted the genetic makeup of the species. These issues have created isolated populations that are vulnerable to local extinctions. A primary goal of the US Fish and Wildlife Service is to develop networks of self-sustaining populations for long-term persistence. If we can abate these threats, one of the West's most iconic sport fishes might just dodge extinction.

LAKE BONNEVILLE

Early Euro-American explorers to the Great Basin quickly noted intriguing benches, or terraces, extending for miles above the shorelines of the Great Salt Lake in what is now Utah. They eventually concluded that these were remnants of an ancient shoreline to a now drained, vast inland lake. Lake Bonneville gets its name from the explorer, mountain man, and fur trapper Captain Benjamin Bonneville (1796–1878). The lake formed during the Pleistocene glaciation when

Figure 2.124. Terraces that formed around the shores of the now-dry Lake Bonneville are visible in this drawing by H. H. Nichols that appeared in Grove Karl Gilbert's 1890 report on the lake. Public domain image.

extraordinary amounts of water flowed into the Great Basin—a region where there is no natural outlet to the sea—and pooled into an enormous lake in what is one of the aridest regions in North America.

The geologist Grove Karl Gilbert published a report on the ancient lake in 1890 that still stands as the authoritative description of this water body.[189] In it, Gilbert noted that the "area of the Bonneville water surface was 19,750 square miles. If its water surface were given a circular shape, its circumference would be 500 miles, but the actual length of coast, exclusive of islands, was 2,550 miles." Furthermore, the maximum lake depth was about 1,050 feet. By comparison, the Pleistocene Lake Bonneville was larger than the modern-day Lake Erie. During its existence, the level of Lake Bonneville periodically rose and fell in sync with Pleistocene climate.

Consequently, as Gilbert observed, the shorelines of the lake were not uniform. "The water rose and fell step by step, but not with equal pace, and at a few stages it lingered much longer than at others, giving its waves time to elaborate records of exceptional prominence." At one point, the lake level rose so high that water escaped from the Great Basin by breaching a divide and briefly draining to the ocean. Gilbert identified the point of discharge as the north end of Cache Valley, at a spot known as Red Rock Pass. The discharging lake entered Marsh Creek valley and merged with the Portneuf River, which in turn flows into the Snake River. This spectacular event, known as the Bonneville Flood, moved 4-foot boulders at approximately 17 feet per second and, at a maximum, discharged water at an estimated 17 million cubic feet per second. Although geologists do not believe the Bonneville Flood created the Snake River Gorge, it certainly modified it in dramatic ways.[190]

With the end of the Pleistocene, the vast glaciers began melting away, and evaporation from Lake Bonneville exceeded the glacial discharge. Lake Bonneville shrank, leaving many isolated lakes. Eventually, most of these dried up, and only a few remain, the largest being the Great Salt Lake.

LEE'S FERRY

This otherwise obscure site in northeastern Arizona became famous in 1922 as the demarcation point where the states apportion water between the upper and lower basins under the Colorado River Compact. Lee's Ferry gets its name from the Mormon colonist John D. Lee who operated a ferry to transport people and goods across the Colorado River. Lee is infamous for his role in helping orchestrate the slaughter of one hundred and twenty non-Mormons in 1857 by Mormon militiamen and Paiute warriors at Meadow Mountain. Brigham Young dispatched Lee to the remote ferry site in the aftermath of the Mountain Meadows Massacre so that he might escape justice for his crimes. Authorities

Figure 2.125. John D. Lee is sitting on his coffin minutes before his execution for the part he played in the Meadow Mountain Massacre. Josiah Francis Gibbs. 1909. *Lights and Shadows of Mormonism*. Salt Lake City, UT: Salt Lake Tribune Publishing Co., p. 234.

eventually apprehended Lee, and in 1877 he was hung for his role in the violence. Lee's Ferry, located just downstream from the Glen Canyon Dam, now serves as the launch site for raft trips down the Grand Canyon.

LEVEE

Natural or human-made barriers or berms along the edge of a stream or river are known as *levees*. Natural levees form as a byproduct of overbank flooding. When rivers flood, the water is often sediment laden. Within the river channel, the velocity of the water is in general relatively high, which keeps the sediment suspended, but when it goes over the bank and onto the floodplain, the velocity decreases. With decreased speed, the energy to keep sediment suspended diminishes, and sediment settles out of the water. Since the heavier and coarser sediment settles first, it usually accumulates near the stream bank. These thicker sediment deposits form levees once the flood water recedes.

People build artificial levees to protect surrounding areas from overbank flooding and to keep the river channel from migrating around within its floodplain. These provide a sense of security for those living behind them. The lessened flood risk enables investments across landscapes that were formerly

Figure 2.126. The concrete levee is flanking the Arkansas River at Pueblo, Colorado. Photo courtesy of the author.

vulnerable to regular flooding. But they are not impervious to failure. Like all interlinked systems, the protection that levees offer is only as reliable as its least robust section. When these levees fail, due to either shoddy construction or simply being overwhelmed by floodwaters, vast areas get inundated and damaged, people die, and investments made over decades may vanish in a single event.

LONDON BRIDGE

Surrounded by a shopping mall, restaurants, and retail establishments, London Bridge serves as a draw for businesses in what would otherwise be an empty expanse of Arizona desert bordering the Colorado River. The story of how an iconic bridge from London, England, ended up across the Atlantic Ocean to span an arm of a reservoir in Arizona is one for the ages.[191]

Lake Havasu itself was formed by the construction of the Parker Dam on the Colorado River from 1934–1938. The lakeshore remained a mostly remote stretch of Arizona desert when, in early 1963, the entrepreneur Robert McCulloch flew over Lake Havasu looking for a place to test outboard engines. It struck McCulloch that the land surrounding Lake Havasu had great potential for a real estate scheme. McCulloch pursued the idea, and that led to the incorporation of Lake Havasu City in September 1963. McCulloch enlisted the developer C. V. Wood (who had designed the Disneyland amusement park in Anaheim, California), and together they founded what would become a thriving community. They

Figure 2.127. London Bridge at Lake Havasu City. Photo courtesy of the author.

purchased land and then marketed their development to sunbirds (people from northern states who wished to escape frigid winters in sunny Arizona).

While McCulloch was seeking marketing ideas that would bring people to Lake Havasu City, he learned that the City of London was looking to replace John Rennie's 1831 "New" London Bridge. It turns out that this stone bridge was slowly sinking into the River Thames and London decided to build a replacement. Rather than simply demolish the scenic stone bridge, London put the historic landmark up for auction. McCulloch put together a bid of nearly $2.5 million US dollars for the structure. As it turns out, there were not many bidders, and McCulloch won. The bridge was dismantled block by block, marked and numbered, and then shipped via the Panama Canal to Long Beach, California. From there, blocks were trucked inland 300 miles to Lake Havasu City. Workers reassembled the bridge, and on October 10, 1971, the completed bridge was dedicated in a ceremony that drew over 50,000 American and British spectators to the Arizona desert. McCulloch's investment panned out, and the London Bridge provided the marketing ointment needed to make Lake Havasu City a thriving community in the Arizona desert.

LOS ANGELES RIVER

Accurately termed "a bleak and dispiriting 51-mile channel," the Los Angeles River winds its way through the heart of one of America's great urban centers.[192] As one historian notes, from "its beginning in the suburbs of the San Fernando Valley to its mouth in the Pacific Ocean, the [Los Angeles River's] bed and banks are almost entirely concrete. Little water flows in its wide channel most of the year, and nearly all of that is treated sewerage and oily street runoff. Chain link fence and barbed wire line the river's fifty-one mile [*sic*] course. Graffiti mark its concrete banks. Discarded sofas, shopping carts, and trash litter its banks."[193] Grim as all this sounds, many Angelinos love this diminished, depleted, and concreted river, as it is the only river accessible to the majority of people in the region.

The Los Angeles River was not always like this. In the nineteenth century, the river flowed freely and had a floodplain dense with willows and sycamores. First, Indigenous people and later Spanish migrants lived along its course. The river supplied all the domestic and agricultural needs for water for the residents. Extensive vineyards produced respectable wines using the river's water. In the years after the Mexican-American War, Americans began settling in the region, especially after the Gold Rush. The population of the small Los Angeles pueblo began to take off. By the end of the nineteenth century, urban development had accelerated, and diversions drained the river to supply the growing city's

Figure 2.128. Hollywood has long used the concreted channel of the Los Angeles River as a backdrop for action scenes, such as in this film still from *Terminator 2: Judgment Day*.

needs. In the mid-1920s, the riverbed was all but dry, and it became a source for aggregate or a dumping ground for trash. Occasional floods would flash through the river's course, and calls increased to contain the flooding. Beginning in the 1930s, the US Corps of Engineers began constructing flood control dams and by 1960 had lined the entire fifty-one-mile river course with concrete. This concrete channel is the state in which the river largely remains today, even though many Los Angelinos call for the river's restoration.

Adding to the Los Angeles River's gritty aura is that the river makes a convenient backdrop for many chase scenes in Hollywood movies and TV shows. These films make the river instantly visible to viewers worldwide, even if they are unsure that what they are seeing actually is a river.

LOSING STREAM

Sometimes the amount of water flowing in a river or stream decreases as it moves downstream. These losing streams are a worldwide phenomenon but are particularly common in the American West. Streamflow loss happens as water infiltrates into the ground. Human-made river diversions also cause decreased

Figure 2.129. Notice the broad snow-covered and sunlit alluvial fans spreading out from the mouth of the canyons in this winter photo of Colorado's Sangre de Cristo Mountains. The snow helps accentuate the ribbons of dark vegetation winding down the slope and eventually disappearing. These ribbons of vegetation trace the course of losing streams. In spring, water flows out of the canyons, but this water is absorbed into the alluvium as it flows downhill. The riparian vegetation is denser near the mouth of the canyon and diminishes along with the water that supports it. Photo courtesy of the author.

flows along a stream and generate geomorphic features that look like natural losing streams. For example, losing streams, both natural and human-made, have channels that become narrower the further downstream you go.

Perhaps one of the classic sites to look for losing streams are desert alluvial fans. One might observe steady streamflow in the canyon above the fan, particularly when the water is flowing over bedrock. However, once the stream exits the canyon and begins flowing over the fan, water is absorbed into the streambed, and the flow diminishes and maybe even wholly dries up.

MEAD, ELWOOD

By the time Elwood Mead (1858–1936) died in office a few short months after the Boulder Dam's dedication (later renamed Hoover Dam), he had devoted his life to the problems of irrigation and irrigators. His career paralleled the rise to maturity of reclamation in the United States. Mead saw reclaiming the

Figure 2.130. A young Elwood Mead in 1889. Image Courtesy of the University of Wyoming, American Heritage Center, Photofile: Mead, Elwood, Negative Number 20356.

desert as more than a civil engineering problem. Instead, he viewed reclamation as encompassing all aspects of putting the land to use and settling it with actual farmers who would work it. During his early career, Mead focused on building the institutions and science of irrigation and ditch management. In the 1880s, he taught mathematics at the Colorado Agricultural College (later Colorado State University). He then became the Wyoming State Engineer, where he worked to incorporate the Prior Appropriation Doctrine into that state's constitution. At mid-career, Mead spent eight years in Australia, working on arid lands and irrigation problems there. Returning from Australia, Mead served as chief of investigation and drainage irrigation for the US Department of Agriculture.

Mead became one of the best-known irrigation experts in the world. In 1924, he was appointed the commissioner for the US Bureau of Reclamation. Ironically, Mead had opposed the Newlands Act, the legislation that led to the Bureau of Reclamation and the construction of the large dams that he supervised. When he died in office in 1936, Mead had directed the construction of the Hoover Dam, a $100 million multipurpose dam that took nearly five thousand men five years to erect. While at the Bureau of Reclamation, he also oversaw the Grand Coulee Dam development in Washington, the Owyhee Dam in Oregon, and many other large water projects. Lake Mead, the reservoir behind Hoover Dam, was named in Mead's honor.[194]

THE MILAGRO BEANFIELD WAR

In John Nichols's 1974 book, *The Milagro Beanfield War*, the fictional New Mexican village of Milagro finds itself at the center of a water development scandal. Part of Nichols's "New Mexico trilogy" (the others being *The Magic Journey* and *The Nirvana Blues*), the book explores the imbroglio between land and water, race, modernity, and tradition in northern New Mexico.[195] Director Robert Redford adapted the novel into a popular movie of the same name in 1988. In *The Milagro Beanfield War*, a predominantly Roman Catholic Hispanic community with a mostly interrelated population struggles for political rights, dignity, and self-reliance.

In Milagro, the reluctant hero, Joe Mondragon, defends his small bean field and his community against ruthless business people and corrupt politicians. A crisis erupts when locals discover that politicians and developers made a back-room deal to usurp the town's water to build a resort. Opposition to the resort comes from a frustrated and out-of-work Mondragon, who kicks down a headgate on his father's farm, releasing water into a dry field. Although the authorities had shut the headgate, Mondragon decides to plant beans and irrigate them.

Figure 2.131. A film still from *The Milagro Beanfield War*. The movie's ensemble cast includes Chick Vennera, Julie Carmen, Ruben Blades, Sônia Braga, Melanie Griffith, John Heard, Daniel Stern, James Gammon, and Christopher Walken.

What ensues is a colorful existential struggle for the soul of a small rural community. Milagro articulates in fictional terms broader issues of ethnic independence, cultural heritage, tradition versus modernity, the future of local agriculture, and the control of water in a gentrifying state.

MISSOURI RIVER

"The Heart of the World" is what the Mandan people call their homeland on the Upper Missouri River. "Desolate, windy, and magnificent" is how Elizabeth Fenn describes the Mandan's ancestral village overlooking the Missouri River that the Mandan occupied for nearly three hundred years.[196] To the Indigenous people who lived along the Missouri River—the Mandan, Arikara, and Hidatsa—this river was and is indeed the center of their universe. And until European diseases and colonization nearly wiped out these people, the Mandan, Arikara, and Hidatsa homeland was anything but desolate. For hundreds of years, their bustling communities stood as gatekeepers to the Missouri River and with it much of the essential trade between eastern and western North America. Within these communities, the Missouri River is sacred. Every aspect of life involved ceremony. There was virtually no distinction between the holy and secular. To trap fish, for example, involved rituals in every part of the process, from making and setting a trap to consuming the catch. With the European incursions, Native control of the Missouri River eventually slipped away.

French explorers Louis Jolliet and Jacques Marquette became the first Europeans to see the Missouri River's mouth during their explorations of the Mississippi River in 1673. By 1705, additional Frenchmen had ascended the lower reaches of the river. But comprehensive exploration by Europeans had to wait until the Louisiana Purchase in 1803. President Thomas Jefferson wanted to understand better what the country had purchased from France and directed Meriwether Lewis to take charge of an expedition to the region. Lewis chose William Clark as his coleader. The famed Lewis and Clark Expedition was born. One of Jefferson's goals for the team was to determine if a practical route to the Pacific existed in its upper reaches. During the nearly two-year, 8,000-mile journey, the intrepid explorers passed through the Mandan, Arikara, and Hidatsa villages to reach the Missouri's source, cross the Continental Divide, and arrive at the Pacific Ocean. This trip opened the region to fur trapping and trading with Indigenous people. By 1819 steamboats began ascending parts of the river. Steamboat traffic peaked on the Missouri River in 1858 with the first railroad construction into the region.

Figure 2.132. Margaret Bourke-White's image of the Fort Peck Dam on the Missouri River adorned the cover of the inaugural issue of *Life Magazine.* According to the Metropolitan Museum of Art, Bourke-White's image "solidified Fort Peck Dam's status as an icon of the machine age. Looking up at giant concrete buttresses and what appear to be crenelated battlements (actually supports for an elevated highway), with small figures on the spillway providing the necessary indication of scale, Bourke-White creates a vivid illustration of the power of technology to dwarf humankind. This robust work presents new, modern monuments as equally impressive as the towering walls of ancient cities." Here, the Missouri River was tamed as other western rivers, placing it squarely at the junction of American power during the twentieth century. Image courtesy of Getty Images.

During the Great Depression, the Public Works Administration began work on Missouri River dams as part of the New Deal to stimulate the American economy. Fort Peck Dam (Montana), the largest of these, constructed by the Corps of Engineers beginning in 1933, created the fifth largest human-made lake in the United States. The Corps of Engineers built four other huge dams on the Missouri, including Garrison (North Dakota), Gavin's Point, Fort Randall, and Oahe (South Dakota). In 1944, the US Congress authorized a flood control and water resource development plan for the Missouri River, including channel deepening, bank stabilization, and extensive levee construction. All of these dams, levees, and channel modifications have fundamentally converted the formerly wild river into one of the most extensively engineered waterways in the nation.[197]

MULHOLLAND, WILLIAM

William Mulholland (1855–1935) migrated to the US in the 1870s as a penniless Irish teenager and rose to become the chief of what was arguably the most influential and powerful municipal water agency in the United States. Mulholland had no formal engineering training, but his voracious appetite for reading hydraulic engineering texts coupled with an aggressive work ethic put him on a path to bring water to a growing Los Angeles. As the legend goes, Mulholland's big break came when, working knee-deep in muck cleaning a ditch, a man on horseback approached and asked him what he was doing. Not bothering to look up, he blurted out, "None of your damn business!" and kept working. Learning later that he had just snubbed his ultimate boss, William Perry, president of the Los Angeles City Water Company, he put down his shovel and went to apologize and submit his resignation. Instead, Perry, impressed by the young man's spunk and hard work, promoted him and gave him a raise.[198]

Mulholland worked his way up the chain, proving himself as a foreman and engineer by completing ever more complicated water projects. Eventually, the City of Los Angeles purchased the Los Angeles City Water Company, leading to the creation of the Los Angeles Water Department (the predecessor to the Los Angeles Department of Water and Power), and with it Mulholland became its first superintendent and chief engineer. One of his critical tasks was to slake his city's voracious thirst for more water. Promoting water projects was how Mulholland became famous, or to others infamous, for he directed the construction and diversion of the Owens River to Los Angeles through the

Figure 2.133. For decades William Mulholland was the power broker who oversaw the dramatic expansion of Los Angeles's water supply. Image courtesy of the LA Public Library.

Los Angeles Aqueduct. By mainly planning the early stages of the project in secret, and because inside supporters purchased land in the San Fernando Valley near where the aqueduct would terminate, farmers and ranchers in the Owens Valley felt that their lifeblood was surreptitiously stolen from them. What unfolded was a decades-long conflict, culminating in the dynamiting of the aqueduct.

Even with the completion of the Los Angeles Aqueduct in 1913, Mulholland sought even more water for the city. Mulholland promoted plans to divert Colorado River water to Los Angeles. Mulholland's final large project was the St. Francis Dam near Saugus. At first, it seemed like a great success, but the dam catastrophically failed on March 12, 1928, when the walls of the nearly full dam collapsed. The total casualties from the disaster will never be known, but at least 424 people died. Mulholland assumed full responsibility for the disaster. He retired soon after. In retirement, Mulholland remained active, consulting on aspects of both the Hoover Dam and the Colorado River Aqueduct.[199]

NATIONAL ENVIRONMENTAL POLICY ACT

Federal legislation, exemplified by the National Environmental Policy Act of 1969 (NEPA) has had widespread influence over water policy and, by extension, the infrastructure constructed in the West. In many ways, NEPA was the culmination of the environmental movement that sprang out of the tumultuous 1960s. Events such as Rachel Carson's 1962 book, *Silent Spring*, which detailed the decimation of bird populations by pesticides, to the June 22, 1969, fire on Ohio's Cuyahoga River led to a groundswell of Americans demanding federal action to protect the environment. NEPA, for the first time, declared national environmental policies. A central provision of NEPA requires the federal government to weigh environmental factors equally with other concerns in rendering federal agencies' decisions. Federal agencies implement NEPA through the preparation of environmental impact statements as a tool to determine how federal projects or decisions may affect the environment. NEPA's power over water projects is pervasive. For example, through the process of writing Denver Water's EIS for its proposed Two Forks Dam, ecologists identified potential wetland impacts. These wetland impacts led to the eventual vetoing of the project by the George H. W. Bush administration.[200]

Figure 2.134. Here is a photograph of President Richard Nixon on January 1, 1970, at the press conference where he signed the National Environmental Policy Act. (Source: image courtesy of Nixon Presidential Library and Museum. Oliver Atkins photographer. Public domain image.)

NONPOINT SOURCE POLLUTION

When pollution comes from a pipe, it is easy to spot and mark those point sources with a big X on a map. Regulators can walk up and measure the pollution and make polluters do something about it. For dispersed sources of pollution, the situation is not so simple. The Clean Water Act targeted discrete pollution sources, and by default, any source of water pollution that one could not attribute to a discrete conveyance became known as nonpoint pollution. In many states, well over 60 percent of all pollution entering streams and lakes come from nonpoint sources. Also, nonpoint pollution, because it is discharged over a vast land area or nonspecific locations, makes it a thornier problem. Nonpoint pollutants often come from sediment, nutrients, organic, and toxic substances originating from land-use activities. Regulators overlooked these sources and postponed to later years' actions to clean these pollutants up. These nonpoint pollutants get into the water via rainwater, snowmelt, or irrigation runoff from farmland, city streets, or suburban lawns. Nonpoint sources contain everything from pesticides, heavy metals, ground-up plastics, and dog poo. Groundwater is particularly vulnerable to nonpoint pollution. Nitrate fertilizers, pesticides, herbicides, feedlot runoff, and other chemicals used by farmers can later seep into the ground. Eventually, all of this pollution gets into the rivers that one's municipality must then clean up before it sends you drinking water through its pipes.[201]

Figure 2.135. A typical storm sewer cap that warns against dumping to protect downstream waterways. Photo courtesy of the author.

OUTLET WORKS

One of the more exhilarating things one can do on a great dam is to stand atop the outlet works when the water is discharging at full force. The structure rumbles with the frightening power of a hundred freight trains and with the unmuffled sound of jet engines on the tarmac. But it is a controlled discharge, engineered for the safe operation of a dam. Engineers design outlet works to move water from the pool above the dam to the river below. Outlet works must have the capability to pass the average and high stream flows that enter the reservoir yet carefully regulate the dam's outflow. Excess water from massive storm or flood events does not go through the outlet works but instead gets diverted around the reservoir through its spillway. Outlet works generally contain one or more pipes or tunnels that pass through the dam embankment. Water entering the outlet works first passes through an intake structure, consisting of a canal or intake tower. Next, a regulating valve controls the water flow through the pipes. Finally, the water exits through pipes first into a stilling basin (to calm the turbulence) or directly into the river below. Water passing through these pipes is under high pressure and is related to water height in the reservoir behind the

Figure 2.136. In 2004, water managers at the Glen Canyon Dam performed a high-flow experiment by fully opening the outlet works to the dam, creating a spectacular plume of water out of the structure. For a sense of the scale, notice the cars on the left of the photo. Courtesy of the USGS, photographer Anne Phillips. Public domain image.

dam. In many dams, the outlet works are attached to hydroelectric turbines to take advantage of the water pressure to generate electricity.

OVER THE RIVER

Over the River was a landscape art installation that the artists Christo (1935–2020) and Jeanne-Claude (1935–2009) had planned to construct along the Arkansas River of Colorado. Their project was a prominent contemporary example demonstrating how artists can provoke local and national audiences into debates over art and the environment. Christo and Jeanne-Claude (they only used their first names) viewed the work as a celebration of Colorado's landscape; however, detractors derided the project as a damaging affront to Colorado's rivers. This clash of values led to years of political and legal wrangling

Figure 2.137. A rendering of Christo's proposed *Over the River* (project for Arkansas River, state of Colorado) landscape art installation. Image courtesy of the estate of Christo V. Javacheff. 2010 pencil, pastel, charcoal, and wax crayon 35.2 x 38.7 cm (13⅞ x 15¼ in.) drawing. Photo by André Grossmann. Copyright 2010.

over a proposed two-week installation. In 2009, Jeanne-Claude passed away, but Christo continued the work. At the cost of over $15 million of his dollars, Christo was within sight of final court approval when he decided in January of 2017 to cancel the project as an act of protest against Donald Trump's inauguration. Had the numerous legal and bureaucratic challenges been resolved, Christo intended to "suspend 5.9 miles of silvery, luminous fabric panels high above the Arkansas River."[202] The project was to have waves of fabric shimmer in the sun and transition in color as the day progressed. Christo decided to abandon the project because the federal government owns the land, and he wished to have nothing to do with a US government controlled by Donald Trump. Even though *Over the River* was never installed, it succeeded as a multiyear performance spectacle that pits art lovers against local activists who see any intrusion onto the riparian corridor as objectionable. His decision to cancel the project placed his art at the center of water and land-use issues and politics in the American West.

OWENS VALLEY

When the water began flowing into the Los Angeles Aqueduct from the Owens River in 1913, Los Angeles secured the supplies necessary to fuel dramatic growth. But those diversions proved a considerable impediment to the desert communities in the Owens Valley that relied on the water for their agricultural economy. Concerns over rural to urban water transfers began here and repeated over the next century and into the present, as large municipalities targeted rural water supplies for acquisition to enable their growth. Viewed by some as a case study in municipal imperialism, Owens Valley became the cautionary tale for any small agricultural community with a large city looming nearby. Initially, Owens Valley residents watched passively as Los Angeles diverted the river's water. However, locals quickly soured with Los Angeles as word of the city's secret land deals and legal moves spearheaded by William Mulholland and his associates to take even more water became widely known. Suspicion and mistrust eventually boiled over into overt anger. Then, on May 21, 1924, that anger turned violent as disgruntled farmers dynamited the Alabama Gates, an important aqueduct facility near Lone Pine.

The controversy over the water transfers soon extended beyond the Owens Valley. A group known as the "San Fernando Syndicate" held inside knowledge of the aqueduct, parlaying that information to purchase land in the San Fernando Valley in anticipation of future water deliveries and real estate riches. Syndicate members included William Mulholland; Fred Eaton, the political mastermind

Figure 2.138. In May 1927, farmers, angry over the water diversions from the Owens Valley to Los Angeles, dynamited sections of the Los Angeles Aqueduct, including the No Name Siphon pictured here. About 300 feet of pipe collapsed as a result of the explosion from dynamite sent into the pressurized line. Courtesy of the Los Angeles Public Library. Herald-Examiner Collection HE Box 1071 HE-001-098 4x5 http://jpg1.lapl.org/pics44/00041991.jpg.

behind the aqueduct project; Harrison Otis of the *Los Angeles Times*; Henry Huntington of the Southern Pacific Railroad; and other well-connected capitalists of Los Angeles.

Unrest stemming from the water transfers continued for years. In 1927 alone, saboteurs damaged the aqueduct at least seven times. The water transfers, Los Angeles's aggressive legal stance, the secrecy and self-dealing, and the violent response by Owens Valley farmers have made the Owens Valley a worldwide symbol of water conflict. Roman Polanski dramatized these events in his Hollywood potboiler, *Chinatown*. Eventually, the violence tapered off with arrests and increased policing. But the bitterness in the valley remained. All the anger welled up again when, in 1976, Los Angeles announced intentions to increase groundwater pumping in the region. In response, saboteurs bombed the Alabama Gates a second time. Not only that, but there was a failed attempt to launch dynamite strapped to an arrow into a fountain honoring William

Mulholland in Los Feliz, California. Owens River water continues to be moved to Los Angeles and remains one of the city's significant sources of supply.[203]

OXBOW

Rivers are dynamic and move about their floodplains. Many alluvial rivers contain broadly looping meanders that migrate over time.[204] When a meander nearly wraps back upon itself, an oxbow that looks like a horseshoe-shaped length of stream channel develops. Once the meander completely closes itself off, by cutting through its bank, the river becomes much shorter. When this happens, the looped portion of the channel is abandoned as part of the river, leaving a lake. Sometimes these lakes are ephemeral or contain mostly wetlands. Eventually, the lake fills in with fine sediment leaving a meander scar on the flood plain.

Figure 2.139. Here are two oxbows along Montana's Milk River. Take a look at the oxbow lake on the left; it is longer and mostly filled with water, indicating that it was relatively recently formed. On the right, the oxbow lake is filled with sediment leaving only a small amount of water; this means that this oxbow is older. The buildings on the upper left suggest the scale. Screen capture image courtesy of Google Earth.

The term itself derives from the resemblance of the lake's shape to that of the U-shaped collar of an ox yoke.[205]

PARKER DAM

Although the Parker Dam is not nearly as famous as the Hoover and Glen Canyon Dams upstream from it, this project has the distinction of diverting the Colorado River water stored in those dams to both California and Arizona. It is also the site of one of the more contentious episodes in western water development. When Parker Dam's construction began in 1934, Arizona governor Benjamin B. Moeur, fearing that his state would lose out on water deliveries from the river, ordered national guard troops to the construction site to halt construction. The soldiers set up machine-gun positions overlooking the construction to make a point. Assisting the Arizona National Guard was a local ferry company that served under the informal name of the Arizona Navy. The stunt succeeded, and

Figure 2.140. Arizona National Guard mobilize, ready to defend their border and protect its water in the so called "Parker Dam War."

Arizona won an injunction stopping work on the dam. Arizona claimed that the structure would allow California to divert more than its share of Colorado River water with the effect of denying Arizona of its allotment under the Colorado River Compact. Eventually work resumed, but the court case the stunt spawned dragged on until 1963 when the US Supreme Court affirmed the division of Lower Basin water with 4.4 million acre-feet going to California, 300,000 acre-feet to Nevada, and 2.8 million to Arizona.[206]

PERENNIAL RIVERS AND STREAMS

"Wonderfully varied though rivers are, each has a physiognomy of its own. Each preserves its characteristics even in the midst of constant diversity. We recognize it, as we recognize a person in different changes of dress. The Ohio has one face, the Hudson another, and each keeps its essential identity. The traveller would not confuse the Rhine with the Danube, or the Nile with the Volga."[207] This is how William Denison Lyman began his treatise on the Columbia River, eloquently explaining how many large perennial rivers take on a "personality" based on how people perceive its ecosystems, geomorphology, and hydrology. In the West, the largest of these perennial rivers, the Columbia, Colorado, Missouri, and Rio Grande, possess unique personalities that influence the ecology, culture, and economies of their respective watersheds.

In the arid West, it is the perennial rivers that best embody the notion of personality. These perennial, or permanently flowing, rivers have outsized importance to all forms of life. Perennial rivers shape the availability and distribution of critical resources, such as (and obviously) water, because they support more significant numbers of plants and animals in their valleys than exist in the surrounding steppes and deserts or along ephemeral watercourses. Said differently, the diversity of flora and fauna along perennial rivers is substantially higher than in other areas in the West. For people, perennial streams have more resources than they require to sustain settlements. Perennial rivers enhance and support trade, travel, and communications. Plus, they possess strategic value for the people who live along the watercourse.

Because of their great importance for life, perennial rivers take on major cultural significance to the people living along them. Indeed, in some cultures, rivers are so critical to a people's identity that they see the river itself as a living entity. The Navajo, for example, assert that the San Juan River is sentient. And in California, the Yurok Tribe recently declared legal rights of personhood for the Klamath River, in an action that was possibly a first for any North American

Figure 2.141. Perennial rivers, such as Arizona's Salt River pictured here during the spring runoff, flow year-round and provide a crucial riparian corridor for wildlife and humans through an otherwise inhospitable terrain. Photo courtesy of Tina Tan.

river.[208] For the Yurok, "rights of personhood" means the river itself has rights equivalent to that of human beings.

Western perennial rivers are simultaneously natural entities that people adapt to and interact with and are cultural entities that people represent and interpret differently. Here, we find nature and culture complexly interwoven.

PLAYA

One can only imagine the disappointment of early travelers when they came upon one of the Great Basin's playa lakes expecting to see their thirst quenched, only to find their hopes dashed in the lake's brackish water. In areas where there is no outlet to the sea, ephemeral lakes known as *playas* may form. Geographers sometimes refer to these features as endorheic, terminal, or sink lakes, and the basin in which they form a playa basin or bolson. One generally encounters playas in arid or semiarid areas. If it is even present, vegetation on the beds of these lakes tends to be sparse or seasonal. Like all lakes, you frequently see fine-grained sediment deposits within the lakebed. But perhaps the signature features of playas are the salt crusts that form as water evaporates from the bed. These lakes tend to have shallow but saline groundwater lying underneath the salt and sediment crust. When the wind picks up over these lakes, it can kick up fine-grained alkaline—and frequently toxic—dust that affects areas downwind.

Figure 2.142. The desiccated bed of Sevier Lake, one of the largest playa lakes in North America, is virtually devoid of vegetation. Although it is technically a lake, it rarely has water these days because irrigation diversions and groundwater pumping from the streams that supply the lake have drastically reduced inflows. So seldom does this lake have water in it that alkali dust storms rise from its bed to occasionally plague Utah's Wasatch Front. Photo courtesy of the author.

In the Great Basin, playa lakes receive less water than in previous centuries as water users divert from streams feeding the playas. This reduced inflow has added to desertification and increased the extent of salt crust formation across the region.[209]

POOL-RIFFLE

Many stream channels contain alternating combinations of pools and riffles that develop either naturally or are constructed to enhance fish habitat. During off-peak discharge, the pools have lower velocity flows than the adjacent riffles. This lower velocity allows fish to rest and feed in the deeper, slower flowing water. It appears that during flood stages, the typically low water velocities in pools and higher water velocities at riffles reverse themselves, which means that sediment scours at high flows from the pools and then deposits in the riffles. Therefore, over time pools and riffles appear to migrate downstream slowly. These features make a particularly valuable habitat for sport fish. Because of

Figure 2.143. Fisherman taking advantage of constructed pool and riffle habitat on Spring Creek, a tributary of the Taylor River, in central Colorado. Photo courtesy of the author.

this, many fishing advocates alter stream channels to enhance the habitat for their prey. Interest in sport-fish habitat has given rise to a consulting and stream modification industry serving the fishing community. Although beneficial to sport fish, smaller endemic species sometimes find themselves at a disadvantage in these environments because there are fewer places for them to take refuge from predator fish.[210]

POWELL, JOHN WESLEY

Few are more famous in the history of western water than John Wesley Powell. Powell (1834–1902), who had slid into obscurity by the mid-twentieth century, was lifted from oblivion with the publication of Wallace Stegner's 1954 biography, *Beyond the Hundredth Meridian: John Wesley Powell and the Second Opening of the West*. In it, Stegner recounted the treacherous expeditions that Powell led down the Colorado River in 1869 and 1871 and described his role influencing western development and American science as the head of the US Geological Survey and its Irrigation Service, and the Smithsonian Institution's Bureau of Ethnology. Before coming to fame for his western exploits, Powell, like many in his generation, served in the Civil War, and rose to the rank of major. At the vicious battle of

Figure 2.144. Doctors amputated John Wesley Powell's right arm at Shiloh during the Civil War. Image courtesy of the US Geological Survey. Smithsonian Institution Archives Neg. No. 10396. Public domain image.

Shiloh in western Tennessee, Powell was struck by a lead minié ball, a wound that necessitated the amputation of his right forearm several days later. At the end of the war, Powell, a self-taught schoolteacher, geologist, and natural historian, began a series of expeditions to the West, first to collect specimens and later to map unexplored rivers. Powell's first expedition down the Colorado in 1869 was primarily a reconnaissance exploit, fraught with privation due to the destruction of one boat and the swamping of others. Several of his men died (whether at the hands of Indigenous people or Mormons remains controversial) after they quit the expedition in the depths of the Grand Canyon. But the journey itself succeeded, propelling Powell on to national fame and the opportunity to repeat the exploit with a far better outfitted second expedition in 1871.

This second trip was a resounding success, and gave Powell the résumé needed to become the second director of the US Geological Survey. In this role, Powell supervised further expeditions and wrote his still influential *Report on the Lands of the Arid Regions of the United States* (1878). In it, Powell distilled into one

Figure 2.145. John Wesley Powell's boat, the Emma Dean, within the Grand Canyon during the 1871 expedition. The boat is named after Powell's cousin whom he married in 1862. Because Powell had only one arm, he could not oar his boat. For this reason, he sat in the chair that is lashed to the top of the boat as he made his way down the river while others in the expedition manned the oars. Courtesy of the US Geological Survey. Public domain image.

document for the first time the critical environmental issues that the United States faced for successfully advancing western settlement. Powell described the West's aridity, warned that the existing system for transferring public land to civilians was poorly adapted to the West's climate, emphasized that rainfall alone could not support farming, and stated that the construction of communal irrigation systems was necessary. Furthermore, although Powell warned of the prospect of disastrous droughts, he also predicted the potential to develop great agricultural districts in the West.

Powell's report was not without serious flaws. In it, he criticized the Indigenous practice of setting forest fires for game management and even suggested that "fires can, then, be very greatly curtailed by the removal of the Indians." Powell pointed to the construction of communal irrigation projects by the Mormons but neglected to mention that the inspiration for these works likely came from observing the Hispanic people's use of acequias in New Mexico. Powell was so assertive with his ideas on western settlement that it eventually undermined political support for his tenure over the US Geological Survey. This led to his resignation and retirement in 1894.[211] Congress ignored many of Powell's recommendations. But Powell lived to see partial vindication when some of his ideas were implemented with the passage of the Reclamation Act of 1902.

Powell also made a mark in academic anthropology as the Director of the US Bureau of Ethnology. His western travels brought him into close contact with many Indigenous tribes and led to a deep appreciation for their culture. Nevertheless, Powell harbored paternalistic—even racist—attitudes toward the tribes and sometimes advocated for removals of Indigenous people. However, under Powell's supervision, the US Bureau of Ethnology published numerous reports describing Indigenous customs, language, and culture.

Like many heroes, the gloss diminishes upon closer examination as we consider Powell's achievements and failures from a contemporary perch. Powell's significance to western water management appears vastly overrated. Many of Powell's land and water management recommendations were rejected or ignored during his lifetime. In death, the public largely forgot Powell until he was resurrected by Wallace Stegner in the 1950s. Were it not for Stegner, the great reservoir in Utah probably would not bear Powell's name. Although Powell served the country as head of the US Geological Survey in Washington, most of the foundational water laws, management practices, and hydraulic systems took root in the western states and territories independent of Washington bureaucrats, including Powell. Yes, Powell made essential contributions in his role. But countless westerners forged ahead writing the laws and building the systems while Powell's reports sat on shelves gathering dust. Indeed, the Bureau of Reclamation was created, in part, based on Powell's encouragement. But many others saw its necessity too. It's quite likely that the broader outlines of western water development would have advanced in much the same way, centered around local initiative and federal support, even if Powell had not written his reports. Many of Powell's ideas seem quaint, even anachronistic today. Perhaps what is most surprising is that we still debate his ideas for managing western lands and water. Yet, for anyone interested in western water, Powell will forever remain the towering figure who commanded that first treacherous journey down the Colorado River.

RAFTING

When Captain John C. Fremont set out on the Weber River toward the Great Salt Lake in an India rubber boat on September 8, 1843, he and his crew became the first people to raft (in a vessel recognizable as a raft) in the western United States. Joining him that day were Christopher (Kit) Carson, Charles Preuss, Baptiste Bernier, and Basil Lajeunesse. Their triangular raft contained three tubes, two of which leaked so badly that the crew had to pump the bellows continually. On the 9th, they reached the delta with the Great Salt Lake and had to drag the boat—sinking up to their knees in mud with every step—eventually reaching saline water deep enough to navigate. By the afternoon, they were paddling out to an island that Fremont named Disappointment Island (for its lack of fertile vegetation; the island was later renamed Fremont Island in honor of the captain). After some recognizance, they returned to shore and continued their journey west.[212]

Of course, Indigenous tribes have a long history of constructing wooden boats in North America dating back thousands of years. Early American riverboats

Figure 2.146. Boaters in a self-bailing raft in Upset Rapid, Grand Canyon. Photo courtesy of the author.

emulated Indigenous and European designs, such as the flat-bottom boat, or pirogue, traders used on the Missouri River. Then, Lewis and Clark employed larger craft, including keelboats with streamlined hulls on their famous 1804 journey. John Wesley Powell led the next great river expedition in the West in 1869. On Powell's journey, they employed oak and pine craft based on whaleboat designs. In these boats and others of the era, the oarsmen faced upriver while rowing as one would do on the ocean. This orientation changed in 1896 when Nathaniel Galloway of Utah turned the seat around to face downriver and better navigate rapids.

Commercial recreational rafting began in 1909 when the first clients paid to run the Grand Canyon. Later, with the airplane industry's advent, companies began mass-producing rubber life rafts for open water rescue. Boaters soon recognized their potential for river use, and a market developed for surplus military life rafts. By the early 1950s, Grand Canyon outfitters began ordering specialized rafts, known as G and later J rigs. All these rafts had one serious setback back, namely, they had closed stitched floors that collected water in each rapid and that the boaters had to drain manually. All this changed in the early 1980s with the introduction of self-bailing rafts with drains in the floor. Self-bailers are now the industry standard and remain the primary whitewater craft in use today.

RAIN FOLLOWS THE PLOW

Among the great myths of western water is the notion that as Euro-Americans moved to the Great Plains and broke up the sod rainfall would increase. Colonists believed, or wished to believe, that as American civilization advanced, land cultivation, tree planting, and the construction of farms and railroads would invariably lead to a permanent increase in precipitation. Boosters and real estate speculators happily promoted this notion to support land sales. The idea had its origins along Colorado's Front Range in the years following the Gold Rush in 1858. Here, colonists noticed that stream flows increased in eastern Colorado in the years after ditches and farms were built further west. Their observations made it into books, newspaper articles, and even government reports. Initially, scientists thought that precipitation increased by some unknown mechanism related to development. The theory that precipitation increased with sod busting was eventually articulated in its most explicit form by a booster from Nebraska, Samuel Aughey, in 1880, and summed up by Charles D. Wilber in the phrase "rain follows the plow."[213] Aughey suggested that once farmers broke the tough prairie sod with their plows, moisture released from the ground would end up in the atmosphere as increased evaporation, that in turn heightened atmospheric

Figure 2.147. Here, farmers use a steam tractor to break up the tough High Plains sod somewhere in South Dakota, ca. 1910. Little did they know that their efforts would eventually contribute to the brutal dust storms of the 1930s. Photograph courtesy of the Library of Congress. Source: https://www.loc.gov/item/2005685068/. Public domain image.

Figure 2.148. In western Nebraska, colonists commonly built their first homes out of sod, such as the Newbecker family shown here near Sargent, NE, in 1866. They sometimes referred to the sod that they used to construct their homes as "Nebraska brick." Many of these farmers subscribed to the idea that "rain follows the plow." Photographer Solomon D. Butcher. Courtesy of the Library of Congress. (Public domain image. Image ID: nbhips 10169)

moisture and consequently rain fall. In territorial Nebraska, people lauded Aughey for his charismatic ideas. However, Aughey found himself routinely dismissed among scientific circles as a "first class charlatan."

Although westerners widely accepted the idea initially, by the late 1880s and early 1890s questions began accumulating. The bitter pill of hard experience, disastrous droughts, and a recognition that increased stream flows came not from sod busting but from hydrologic changes related to ditch digging convinced people of the idea's nonsense. By the late nineteenth century, some of the greatest scientific minds in the country evaluated and summarily rejected the idea. Among them was John Wesley Powell who concluded that the increase of rainfall by building farms or railroads "has not been shown." Ultimately the rain-follows-the-plow notion led to grief among the many colonists who trusted in it. Blind belief in its veracity, coupled with poor tilling practices, excessive planting, and overgrazing, led to widespread desertification and the Dust Bowl catastrophe of the 1930s. With the idea finally put to rest as useless pseudoscience, colonists could focus on improving farming and ranching practices rather than a hoped-for relief from the sky that never would arrive.[214]

RAINWATER HARVESTING

Rainwater harvesting started as a counterculture back-to-nature strategy for people living off the grid. But over the years, rainwater harvesting, sometimes called rooftop water collection, has evolved into something far more mainstream. Rainwater harvesting systems vary from rain barrels that collect water on residential buildings to large umbrella-like arrays. These systems are legal in all western states but several states have specific rules governing their installation. After many years of debate, Colorado passed laws allowing rainwater collection but only if the homeowner already has a well permit for the property. Opposition to rainwater harvesting in western states comes from the theory that if too many homeowners were to collect rainwater, flows to streams would diminish and harm senior water rights holders. This argument seems ludicrous to someone receiving a few barrels of rainwater when they see someone flood irrigating a field. However, many farmers perceive rainwater harvesting as an incremental encroachment on their water rights. Although the number of people installing rainwater harvesting equipment will likely grow, it is doubtful that these systems will measurably harm downstream water rights owners in a meaningful way.[215]

Figure 2.149. This is a typical rain collection barrel and was part of the US Department of Agriculture, People's Garden Initiative. Because most rooftop collection systems are this size, it is quite understandable that homeowners think it is silly for agricultural water interests to oppose the installation of these facilities. Still, some farmers remain concerned that if many of these systems are installed, they might see reduced stream flows on the streams where they divert water. This shed and rain barrel was displayed at the 2012 Smithsonian Folklife Festival, on the National Mall, in Washington, DC. Image courtesy of the US Department of Agriculture, public domain image.

RAPIDS

The great rapids of the American West, Lava and Crystal in the Grand Canyon, Skull in Westwater Canyon, Ladle on the Selway River, Velvet Falls on the Middle Fork of the Salmon, to name a few, all share some common characteristics. Foremost is the first time any rafter who scouts these rapids will usually experience a deep sense of angst about what they are about to encounter when looking at the maelstrom boiling below them. I certainly did. When I first saw Lava and realized what I was about to oar through, I felt like retching. I had seen a YouTube video of a botched run, with the raft frame torn from the tubes. At Lava, the difference between a clean run and flipping or worse meant following a vague line of foam. If I drifted an oar length to the right or left before the drop, the result could have been ugly. We got through upright but soaked: happily, it was a great run.

With all of these rapids, the river slope increases in gradient and the channel narrows. Exposed bedrock and debris or boulders clog the channel, making navigation difficult or hazardous. Here, the water velocity and turbulence increase, sometimes dramatically, as the water flows over rocks or gravel bars. Rapids vary from river to river depending on the local geology, geomorphology, and hydrology. It is often said that boaters never run the same rapid twice. This axiom refers to how rapids change at differing flows. At flood flows a rapid can look vastly different than at low flows. But don't assume that flood flows mean more difficult rapids: sometimes at high flows the rapid washes entirely out leaving no hint of what is present at lower flows. Conversely, there are rapids that become progressively more difficult—to the point of actual danger—at higher flows. A boater should carefully consider his or her skill level and the flow of the river before committing to any river trip. A well-prepared boater can have great fun. An ill-prepared boater may face disaster.

Boaters and other river enthusiasts use a standard scale to rate the difficulty for running rapids. This scale allows someone who has not floated a particular river to have an approximate understanding of the technical complexity and skill needed to make it safely down the channel. In the United States, the most common rapid classification in use rates rapids on a scale of increasing difficulty ranging from I to V (it's customary to use Roman numerals). For good measure, an additional difficulty designation of VI indicates that boaters consider a particular rapid unnavigable or extraordinarily dangerous. However, North America's most magnificent whitewater river, the Grand Canyon of the Colorado, because of its exceptional size and unique difficulty, alone utilizes a rating scale of 1 to 10.

Figure 2.150. On a large river, such as Idaho's Snake River, an unnamed Class I rapid like this one may look more challenging to run than it actually is. Photo courtesy of the author.

Since boaters mostly use the I to V scale, I'll defer to the experts at the American Whitewater Association for their rating descriptions:[216]

> **CLASS I: EASY**—Fast moving water containing riffles and small waves. Few obstructions, all obvious and easily missed with little training. The risk to swimmers is slight; self-rescue is easy.
>
> **CLASS II: NOVICE**—Straightforward rapids with wide, clear channels which are evident without scouting. Occasional maneuvering may be required, but rocks and medium-sized waves are easily avoided by trained paddlers. Swimmers are seldom injured and group assistance, while helpful, is rarely needed. Rapids that are at the upper end of this difficulty range are designated Class II +.
>
> **CLASS III: INTERMEDIATE**—Rapids with moderate, irregular waves which may be difficult to avoid and which can swamp an open canoe. Complex maneuvers in fast current and reasonable boat control in tight passages or around ledges are often required; large waves or strainers may be present

Figure 2.151. Canoeists skillfully navigating Government Rapid on Utah's San Juan River. At this level, the rapid is rated a Class II. At higher flows, when the velocity makes the water more turbulent and the rocks more challenging to avoid, its difficulty increases to a Class III. Photo courtesy of the author.

but are easily avoided. Powerful eddies and powerful current effects can be found, particularly on large-volume rivers. Scouting is advisable for inexperienced parties. Injuries while swimming are rare; self-rescue is usually easy but group assistance may be required to avoid long swims. Rapids that are at the lower or upper end of this difficulty range are designated Class III- or Class III+ respectively.

Figure 2.152. A rafter scouts Joe Hutch rapid, on the Green River in Desolation Canyon, UT. About two weeks before this photo panorama was made in early September 2008, a flash flood from Joe Hutch Creek deposited the boulders and gravel that gave this rapid its Class III difficulty rating. Photo courtesy of the author.

CLASS IV: ADVANCED—Intense, powerful but predictable rapids requiring precise boat handling in turbulent water. Depending on the character of the river, it may feature large, unavoidable waves and holes or constricted passages demanding fast maneuvers under pressure. Boaters may need a quick, reliable eddy turn to initiate maneuvers, scout rapids, or rest. Rapids may require "must make" moves above dangerous hazards. Scouting may be necessary the first time down. Risk of injury to swimmers is moderate to high, and water conditions may make self-rescue difficult. Group assistance for rescue is often essential but requires practiced skills. For kayakers, a strong roll is highly recommended. Rapids that are at the lower or upper end of this difficulty range are designated Class IV- or Class IV+ respectively.

Figure 2.153. Boaters scouting Warm Springs rapid on the Yampa River, a Class IV rapid, at a flow of about 4000 cfs. Photo courtesy of the author.

CLASS V: EXPERT—Extremely long, obstructed, or very violent rapids which expose a paddler to added risk. Drops may contain large, unavoidable waves and holes or steep, congested chutes with complex, demanding routes. Rapids may continue for long distances between pools, requiring a high level of fitness. What eddies exist may be small, turbulent, or difficult to reach. At the upper end of the scale, several of these factors may be combined. Scouting is recommended but may be difficult. Swims are dangerous, and rescue is often difficult even for experts. Proper equipment, extensive experience, and practiced rescue skills are essential. Because of the broad range of difficulty that exists beyond Class IV, Class V is an open-ended, multiple-level scale designated by class 5.0, 5.1, 5.2, etc. Each of these levels is an order of magnitude more difficult than the last. That is, going from

Figure 2.154. Lava Falls remains the crux of any float trip through the Grand Canyon of the Colorado River. On the Grand Canyon scale of difficulty, it is considered a Class 10 rapid, and under the American Whitewater Association scale of North American Rivers it rates a Class V. Here, a rafter manages to stay afloat as a huge lateral wave in Lava Falls engulfs his boat. Photo courtesy of the author.

Figure 2.255. Selway Falls (shown in this panoramic composite image), located just below the final takeout for the famed stretch of the Selway River known for its many advanced rapids, is considered all but unrunnable except by a few extreme kayakers who would brave its potentially deadly drops. A massive boulder in the center of the falls occludes the passage, making this a Class VI rapid. Photo courtesy of the author.

Class 5.0 to Class 5.1 is a similar order of magnitude as increasing from Class IV to Class V.

CLASS VI—EXTREME OR EXPLORATORY—Runs of this classification are rarely attempted and often exemplify the extremes of difficulty, unpredictability and danger. The consequences of errors are severe and rescue may be impossible. For teams of experts only, at favorable water levels, after close personal inspection and taking all precautions. After a Class VI rapid has been run many times, its rating may be changed to an appropriate Class 5.x rating.

RECLAMATION

In nineteenth-century America, the idea of "reclaiming" desolate western landscapes took on an evangelical aura. Indeed, the roots of reclamation practice incorporated biblical notions of men entering the godless wilderness to convert unusable pagan lands into healthy, holy, and wholesome farms suitable for good Christian families. Euro-American colonists moving west did not seem too worried about the moral contradictions that their reclaimed lands had only recently been appropriated from Indigenous people through ethnic cleansing and genocide. Still, farmers were initially intent on building small ditches so that they might irrigate farm holdings. But within a few years of the first diversions, entrepreneurs realized that by building ditches they could dramatically increase the value of otherwise dry lands. And, once the railroads acquired vast tracts of western lands through government grants (of former Indigenous land) made as incentives to build railroads, these new landowners began seeking ways to cash in on their land bounty. And what better way to increase the value of these lands but to bring water to them and sell the properties to prospective farmers at a high profit. So, the evangelical zeal to irrigate was tightly bound with an even greater biblical desire for profit. Soon, private corporations began raising capital to reclaim western lands. At first, they were hugely successful. But as the cost to build ever bigger ditches, dams, and diversions multiplied, the profitability of newer projects became constrained. With that, water developers began lobbying states and the federal government to support reclamation projects as a public good that would further western development.

Eventually Congress took up the banner of reclamation and passed the Newlands Act, a bill that Theodore Roosevelt signed on June 17, 1902.[217] This act ushered in the age of federal reclamation projects in the US West. The act established the Reclamation Service (the precursor to the Bureau of Reclamation) and a revolving fund of revenue from public land sales to build reclamation works. Farmers built distribution canals and prepared the land for cultivation.

Figure 2.156. After Congress passed the Newlands Act, the Reclamation Service wasted little time in building several large reservoirs. The Reclamation Service quickly established itself as the nation's chief expert on dam construction and irrigation. In the early twentieth century, the public considered large dams a technological marvel. Roosevelt Dam, shown here, located on the Salt River northeast of Phoenix, was among the first projects built by the Reclamation Service. Public domain image.

Initially, water users were required to repay the construction costs over ten years, thus replenishing the fund. In many instances, the Bureau of Reclamation extended the repayment deadline, sometimes several times. The Newlands Act also allowed for homestead entry on public land and set the amount of land any one landowner could irrigate at 160 acres. However, many farmers avoided this limitation by having deeds issued to multiple family members. Eventually, revisions to the Reclamation Act increased the amount of land any single farmer could own to 960 acres. Although Congress designed the Newlands Act to assist small farms, changes to reclamation law have allowed large corporate farms to gain significant amounts of federally subsidized water over the last century.

Reclamation certainly transformed the American West. Farmers and ranchers now irrigate millions of acres of formerly dry land. Dams span dozens of rivers, often taking the lion's share of water in them. Thriving cities sprawl across the former Great American Desert. Reclamation has wrought irreversible changes to the West's environment, including the destruction of habitat, species decline, the introduction of exotic species, and interrupted migration routes, to name a few. How one views all this change is for the reader to decide. Do you possess nostalgia for a time of abundant wildlife and free-flowing rivers? Perhaps you see verdant farms as the highest form of culture? Were these transformations good, bad, or something else? In reclamation, Americans reimagined what the West could be and changed the land to conform to that vision. As new generations come along, they too will leave their imprint on the land as they reimagine reclamation in the twenty-first century.

RECREATION

Some of the most diverse uses of water in the American West involve recreation. River use is part of the so-called "experience economy" and is a significant factor in adventure tourism in the West. In particular, fishing and boating (both river and lake) constitute an enormous economic force in many states. In 2011, for example, nearly 33 million anglers participated in over 443,000 fishing trips that generated more than $40 billion in retail sales.[218] Water-based recreation often adds significant amounts of money to local economies. In just two river basins in Idaho and Wyoming alone, fishing-related employment contributed an estimated $12–29 million to county income and created 341–851 jobs.[219] Direct spending on fishing and boating supplies, guide services, rafting companies, fishing and hunting licenses, along with entrance fees to parks, represent considerable direct contributions to local economies. Similar benefits accrue from recreational rafting. For example, as far back as 1992, about 4,500 people annually floated Idaho's Middle Fork of the Salmon River. Those trips added about $5,830,000 in revenue and 238 local jobs to this area of Idaho. Likewise, Colorado's nonmotorized boating economy is enormous, with boaters spending about $1.3 billion annually in the state.[220] The economic impacts of river-based recreation will likely continue to

Figure 2.157. An angler showing off her catch on the Main Salmon River in Idaho. Photo courtesy of the author.

grow. Not only this, but the indirect spending adding to local economies includes food, lodging, shopping, and gas. These recreational benefits create a strong argument to increase rivers' access and enact laws and policies to allow boaters to use public waters freely. Conversely, we can love these waterways to death, as numerous recreationists vie for access to a limited resource. The challenge moving forward will be to strike appropriate balances between use and river protection, lest both the river environment and user suffer.

RETURN FLOWS AND TAILWATER

People use water to irrigate crops, provide water to stock, satisfy industrial uses, and so on. But not all of this water is entirely consumed by the plants, animals, or industrial processes for which the water was diverted. Some water is needed to saturate the ditch itself, additional water might saturate the irrigated field, and some water might pass into the ground or run off the field and flow back to the stream. All this water that is diverted and not consumed is referred to as *return flow*.

Tailwater is one type of return flow and refers to the water that appears directly downstream from some sort of water facility or use. This can be the

Figure 2.158. This nondescript stream at the western end of Colorado's Grand Valley is dominated by turbid agricultural return flows. If it were not for these return flows, this stream would likely be dry for large portions of the year. Photo courtesy of the author.

water at the downstream end of a dam, irrigated field, or pipe. Managing the quality or quantity of this water is often quite consequential. For example, the temperature of water discharged from a dam may have significant impact on fish or other organisms living downstream from the structure. Similarly, tailwater exiting an irrigated field may contain high levels of dissolved solids, nutrients, pesticides, or other chemicals that can adversely impact stream quality or other water users.

To some, return flows and tailwater are seen as a type of waste, to others, it is a normal consequence of water use. How you view it really depends on what kind of use is essential to you. If your primary goal is to provide water for agriculture, return flows are simply part of the practice. If you want water to remain in the stream to meet environmental needs like fish habitat, it might be viewed as wasteful. Complicating matters is that some return flows support wetlands or recharge aquifers. Plus, with water, one person's waste is another person's supply. Likewise, excessive return flows may indeed mean that too much water is being diverted. For water rights holders, possible changes in return flow patterns are fraught with risk as this may alter supplies to downstream water users. What all this means is that when someone wishes to modify a water right, return flow patterns and amounts are often carefully monitored and any proposed changes to them may entail a detailed engineering and legal review.

RINCON

In the southwest United States, people sometimes refer to an abandoned meander as a *rincon*. In canyon country, you may spot a rincon where a river flowing in an incised meander has broken through a narrow point to leave an isolated mesa surrounded by the former river floodplain. Rincon, like many other terms describing landscape features in the West, comes from the Spanish word of the same meaning.

On a recent raft trip down the Green River in Desolation Canyon, my companions and I hiked around the rincon at Three Canyon. The hike consisted of

Figure 2.159. Here is a rincon located along Stillwater Canyon of the Green River in Canyonlands National Park. The Green River used to flow around the isolated butte on the far side of the river. Initially, the Green River eroded downward creating an incised meander. Eventually, the rincon formed when meander became narrower and narrower at the neck and finally connected to leave the stranded butte. When the meander cut itself off, it shorted the distance the river flows, and left the butte unattached to any of the surrounding cliffs. The Green River's former floodplain surrounds the butte. Screen capture of image courtesy of Google Earth.

a circular stroll in a grassy park-like valley bottom with towering desert cliffs surrounding us. In the center of the valley stood a pagoda-like butte as high as the cliff tops in the distance. Under the late-August sun, the waving grasses were light brown and desiccated. Where erosion had broken through the grass and topsoil, stream cobbles from the Green River were strewn about. Had we not known that the Green River had once flowed across this valley floor and eventually broken through to shorten its long course to the sea, this valley would have presented an odd enigma. But because we knew this was an abandoned meander, we could walk confidently forward and know that without turning around we would reach our starting point.

RIO BUENAVENTURA

During the European exploration of the North American continent, an overarching objective was to find a navigable route from the Atlantic to the Pacific. Even when it became apparent that an all-water passage did not exist, explorers sought rivers that might shorten the journey across the vast continent. Filling the void on the maps of the eighteenth and early nineteenth centuries, cartographers placed a series of mythical rivers between the Rocky Mountains and Pacific. After all, they thought, how could such a vast area exist without one or more significant rivers draining it? Mapmakers even gave the biggest one a name, the Rio Buenaventura (it was occasionally called the San Buenaventura), and then explorers went out to find it.

With the Dominguez-Escalante expedition of 1776, Spaniards made the first comprehensive European traverse of the Colorado Plateau. Their maps from that expedition were generally very accurate. Still, they erred in suggesting that a large, possibly navigable, river exited the Great Salt Lake in what is now Utah and that the Green River might drain westward as well. Maps derived from that expedition depicted the Rio Buenaventura, and a legendary river was born. By the turn of the nineteenth century, Alexander von Humboldt, the famed geographer, completed a map of New Spain that included the Rio Buenaventura. When Humboldt visited Washington in 1804, he gave a copy of his map to Thomas Jefferson, which led Zebulon M. Pike to draw the river on his charts. Over time, cartographers proposed various routes for the river, and many believed it drained toward San Francisco Bay. With the mountain men's explorations, including Ashley, Smith, Walker, and Ogden, the probable existence of the river began to recede. Finally, futile searching during John C. Fremont's second

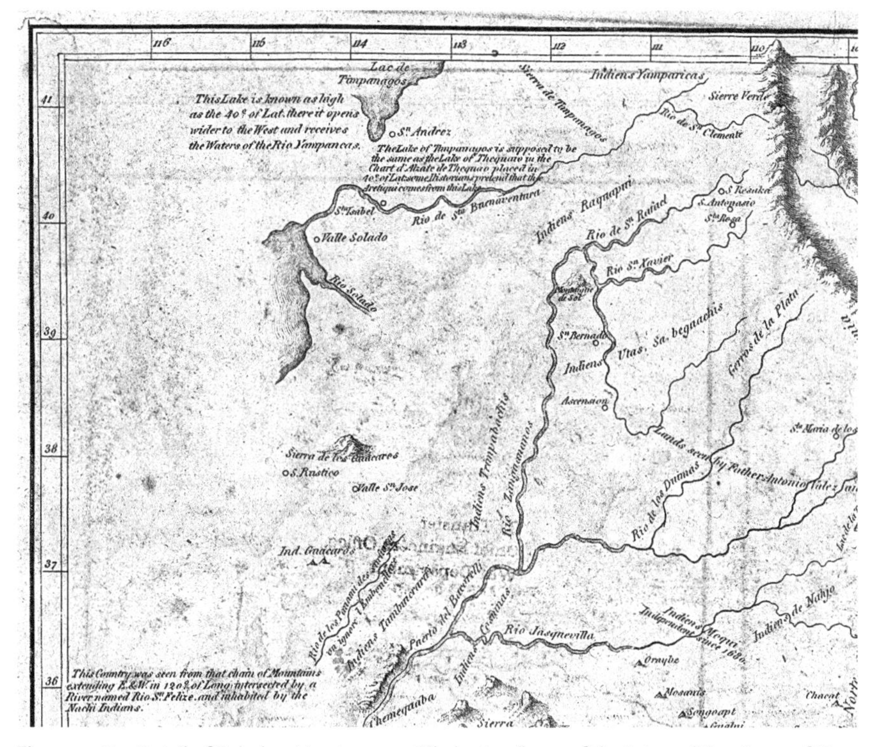

Figure 2.160. Detail of Zebulon Montgomery Pike's 1807 "map of the Internal Provinces of New Spain" showing the "Rio de la Buenaventura" in what is now Utah and Nevada.

expedition of 1844 dispelled the notion that the river existed, and the fictitious waterway disappeared from maps.[221]

RIO GRANDE

An international river, the Rio Grande has its headwaters in Colorado, traverses the entire length of New Mexico, and then forms the boundary between Texas and the Republic of Mexico.[222] Beyond its obvious economic importance as a source of water in its vast watershed, the Rio Grande is of immense symbolic significance for its role in the expansion of America in the nineteenth century and as the boundary between two nations in the twentieth and twenty-first.

The Rio Grande played a vital role in the expansion of empires through time. With Coronado's descent of the river in 1540, New Mexico and southern Colorado became an outpost of the Spanish Empire. Then in 1821, with Mexico's

Figure 2.161. This is the Rio Grande gorge north of Taos, New Mexico. To rafters and kayakers, this area is known as the Taos Box and contains many challenging rapids. Here in the depths of the gorge, surrounded by lava flows of the Rio Grande Rift, the river looks like a thin ribbon that belies its importance as one of the major rivers in the American West. Photo courtesy of the author.

independence from Spain, the river played a central role in the contested boundary between the United States and Mexico. Although Texas declared independence from Mexico in 1836, it was always clear that it was only a matter of time before Texas would become a part of the United States. With the Mexican-American War and the Treaty of Guadalupe Hidalgo that ended it, the Rio Grande was established as the formal boundary between the two nations for much of the region east of the Continental Divide. That same treaty formalized the transition of the Mexican state of *Nuevo Mexico* into the American territory of New Mexico.

During the late twentieth century and into the twenty-first, the river has come to symbolize the difficulty of imposing international borders across what were formerly well traveled paths dating back into prehistory. In addition, photographs, stories, and Hollywood movies tend to inflate (and reify) the image of the river as an unlawful and dangerous passage. For these reasons, along with the Rio Grande's historical and cultural legacy, this river towers over other rivers of similar volume or length as an icon for water in the West.

RIVER ACCESS

Each state in the West has different rules and laws for accessing rivers and streams. What laws work for boaters may not work for fishermen, and what is suitable for a landowner may not be good for the local citizenry or recreation industry. To untangle some of this, I have prepared a condensed summary of access rules across the West. But here is a caveat: this is only a summary and not meant to be a legally definitive compilation. The rules seem to change all the time, so check with local authorities before launching, fishing, or doing something that might impact an adjacent landowner or the future use by other river users. And whatever you decide to do, please remember that when you float past private land, be respectful of their privacy and solitude. If you encounter a hostile landowner, keep in mind that if you remain polite and calm, the high ground is yours, and the encounter is less likely to end up with a call to the sheriff.

Figure 2.162. Each western state has a diverse set of rules guiding access to local rivers and streams. Check local authorities before floating, fishing, tubing, or hiking along stream banks to determine if you can legally access that reach. Shown in this photo is Boulder Creek, CO. Photo courtesy of the author.

When the colonists came West, they rejected English common law that held sway in the eastern United States. In its place, they adopted the Prior Appropriation Doctrine to guide access to water and its use. In it, the courts recognized that water was necessary for life and the development of this region. It allowed access to water for beneficial purposes, recognized the right for navigation, and even allowed for the condemnation of rights of way across private and public land to access water. But the courts and legislatures do not always see recreational access as being similar to access they have granted to diverters. And as each state developed, their legislatures and courts have gone in divergent directions, making laws and interpretations on river access that vary widely across the region.

One more consideration is worth bearing in mind: the land on one or both sides of that river you float by might be private, state, federal, or Indigenous land. Just because you are floating does not necessarily mean you can land, camp, or hike on the land surrounding the river. And what one does will undoubtedly affect other users. In 2018, for example, the Ute Indian Tribe of the Uintah and Ouray Reservation (Ute Tribe) closed the entire length of Desolation/Gray Canyons (river left or east side) on the Green River within their Reservation. This prohibition includes all camping and hiking. Apparently, two private river trips removed elk and deer antlers from the Reservation, stole some artifacts, damaged some historic structures, and failed to pay the required permit fees to use the Reservation. Behavior like this would sour any private landowner toward granting the public access to their land.

Here's the summary for river access in the West:[223]

Arizona

Arizona has some of the least boater friendly river access laws in the West. In 1992, Arizona empowered the Arizona Navigable Stream Adjudication Commission to determine the navigability of state rivers. This commission decided the Colorado River was the single navigable waterway in the state; meaning, that this is the one place that the public clearly has recreation rights. Under all other Arizona rivers, the title to streambeds remains either with the federal government or private individuals, depending upon who owned the land at Arizona's statehood. This means that the public may recreate on waters above federally owned streambeds but not above privately owned streambeds without the landowner's permission.

California

As long as a someone can float a boat on a river for most of the year, California recognizes the public has the right to use the stream. Public use may include recreational boating, fishing, swimming, hunting, etcetera, up to the high-water mark. Unlike other states, California courts have adopted a state test for determining which streams are subject to a public right of navigation. Streams subject to this right include those waters that are navigable at present by any watercraft, including small recreational or pleasure craft propelled by motor or by oars, such as canoes, rafts, or kayaks. However, to be considered navigable, the stream must be suitable for public use, and the state determines this on a case-by-case basis.[224]

In California, the state acquired title to its navigable waters when it was admitted to the union. California holds navigable waters and the streambeds beneath in trust for the benefit of the people. Additionally, California's constitution allows the public to use all navigable waters in the state and directs the legislature to give that provision the most liberal construction.[225] It doesn't matter whether the streambed of a navigable river is publicly or privately owned; an easement for public navigation and related uses (boating, fishing, swimming, hunting, and other recreational uses) is recognized. In California, the easement exists up to the high-water mark. California has not determined whether portaging and scouting above the high-water mark is allowed. Furthermore, there is no right to trespass across private property to access navigable waters. But where a public road or bridge easement crosses private property at a waterway, the state may consider this as lawful access to the waterway.

Colorado

A 1979 Colorado Supreme Court decision found that a constitutional provision declaring streams to be publicly owned did not grant public access for recreational purposes when the water flowed through private property. It also concluded that the adjacent landowner may own the land under the river and recognized that the water flowing in the stream is the state's property. Although this court decision was somewhat inconclusive, a follow-up State Attorney General's opinion interpreting that ruling concluded that floating through private property does not subject rafters to criminal prosecution. What remains unsettled in Colorado is whether a rafter or fisherman can be found guilty of civil trespass for floating across private land. Furthermore, no rulings or legislation resolve whether river users may legally portage across private land around obstructions (such as diversions) in the stream.[226]

Idaho

Idaho has some of the most boater-friendly river access laws in the United States, even declaring that Idaho rivers are "highways for recreation."[227] Basically, streams that can be floated by a kayak are open to the public for any recreational purpose. Idaho also expressly permits portaging over private land and around irrigation dams or other obstructions that interfere with the stream's navigability. The only restriction is that the boater must reenter the stream immediately below the obstacle at the nearest point where it is safe to do so. Boaters may also lawfully scout within the ordinary high-water mark of a river. Although boaters may use the stream up to the high-water mark, one recent Idaho Supreme Court case placed some limits to this rule by eliminating public access to a beach that has been used by swimmers for a century.[228]

Montana

Montana's constitution states that all waters of the state are owned by the state for the use of its people. Montana's courts interpret this constitutional provision as establishing a public trust over the state waters.[229] Additionally the Montana legislature adopted a statutory recreational use test to define navigability for purposes of establishing the public trust. This navigability test states that "all surface waters that are capable of recreational use may be so used by the public without regard to the ownership of the land underlying the waters."[230] Montana also recognizes that the use of a water body is permitted up to the high-water mark

Montana has further expanded the public's recreation rights over private streambeds. In a 1988 decision, Montana's Supreme Court ruled that the public could use county road bridges to access nonnavigable streams flowing through private property. More recently, in a 2008 decision, the Montana Supreme Court ruled that a reclaimed irrigation ditch supplied exclusively by headgate and irrigation return flows qualified as a natural, perennial flowing stream open to public recreational use.[231]

Nevada

Nevada uses a federal commercial navigability test, meaning that those streams capable of being used or that have been used as a highway of commerce (e.g., floating logs) can be floated, and perhaps fished, recreationally.

On these rivers, the title to the bed of the rivers is held by the state. It does not matter if the waterway is interrupted by occasional natural obstructions or portages, and it doesn't matter if the stream is not runnable all year for it to be considered navigable.[232] Nevada has not determined if it is legal to portage obstructions. In Nevada, statute or by case law have declared as navigable to some degree the Colorado, Virgin, and Carson Rivers.[233] If a river or stream passes over private property on a stream other than these, and the landowner has posted a no trespassing sign, a boater of fisherman may be cited for trespassing for entering that stream reach.

New Mexico

In New Mexico, the title to land under streambeds may be held by private entities. Water users maintain only a usufructuary right (right by use) in the water, and the state owns the title to the water. This includes all unappropriated water from every natural stream, perennial or torrential. New Mexico courts interpret this to mean that the public has the right to use water in streams and rivers for recreational purposes, subject to appropriators' right to remove water from the stream. In other words, in New Mexico, you can float any stream for which there is legal access.

That said, in 2015, the New Mexico Legislature rushed through a hastily introduced measure (Senate Bill 226) without significant hearings of public input on the next to last day of its 60-day session. The governor promptly signed the bill stating that recreational users can't walk or wade onto private property through "non-navigable public water" without written permission from the adjacent landowner. As soon as this legislation passed, some landowners began erected fencing or wires across various stream reaches to enforce what they believed was their newfound right to exclude the public. However, in 2022, the New Mexico Supreme Court ruled unanimously in favor of recreation groups who sued over access to waterways running through private property under Senate Bill 226. The Supreme Court affirmed New Mexicans' constitutional right to paddle all of New Mexico's rivers and streams. The court affirmed that water in New Mexico's rivers and streams belongs to the public. The court determined that paddlers have the right to travel these waterways unimpeded by private landowners. Moreover, the New Mexico Supreme Court held that rivers are not for the exclusive benefit of private landowners and that rivers and public recreational users of the public water have an access easement.[234]

No case law or statutes allow river users to portage across private land around obstructions (such as diversions) on New Mexico streams. In New Mexico, the public doctrine gives the public the right to use the zone between the high and low water marks, but above that zone, a boater or fisherman needs permission from the landowner. And the landowner must post no trespassing signs on their property at all vehicular access entry points. Therefore, this aspect of river use remains somewhat unsettled in New Mexico.

Oregon

Oregon has somewhat complicated river access standards and uses a "public use navigable" test to make determinations. What this means is that at the time of statehood, if an Oregon waterway was capable of being used as a highway of commerce for trade and travel using a "customary mode of water transportation," the stream was considered navigable.[235] Oregon's State Land Board has the legal authority to make these navigability determinations. The State Land Board also maintains a list of streams that are navigable under the statutes. Even so, the navigable status of many Oregon streams remains undetermined. On these navigable rivers, Oregon owns the bed of the river, and the public has a right to float, wade, and fish up to the ordinary high-water mark.

On smaller, nonnavigable streams, the public may also have a right to float, fish, and swim, but the law is unclear whether the public can touch the streambed or banks while floating. Furthermore, no formal mechanism exists to determine whether a smaller waterway is subject to the public use doctrine that would allow floating rights. However, Oregon's common law defines navigable waters as a "floatage easement" or "public use" doctrine. Under this doctrine, the public has the right to use a waterway when the bed is privately owned, provided that the watercourse has the capacity, in terms of length, width, and depth, to enable boats to make successful progress through the reach, even if it is for recreational use. Even so, the Oregon courts have not ruled whether the floatage easement includes other rights incidental to boating, such as wading, fishing, and portaging, that involve the use of privately owned stream banks. Consequently, many uncertainties remain when accessing the smaller streams in this state.

Utah

In Utah, the public owns the water and has the right to use any surface water where legal access exists. This allows the public to use the surface water for recreation, including boating. Utah law has not adopted a test of navigability for rivers, nor has it defined whether streambeds of nonnavigable waters can be used or touched by boaters. As a matter of state policy, Utah recognizes a public interest in the use of state waters for recreational purposes.

Utah also recognizes that a public easement over the waters of the state exists, regardless of who owns the water, and the public is not trespassing when accessing these waters. This public easement includes the right to use the water to float leisure craft, hunt, fish, and participate in any lawful activity when utilizing the water. On the other hand, Utah's public easement is not well defined. For example, the Utah Supreme Court has not determined whether the easement includes use of the streambed including the incidental contact of the streambed by an oar or foot. Furthermore, the state has not defined the right of portage in nonnavigable rivers, which means that river users are taking a risk of being charged with trespass should they portage around a diversion, even if it contains deadly hydraulics or obstructions.[236]

Washington

Not to be outdone by any other state in designing odd criteria for floating and navigability, Washington bases its public access to use streams on whether it can float a "bolt of shingles" during high flows. No doubt, some anonymous technocrat spent many hours coming up with a standard definition of a "bolt of shingles" only to have his work second-guessed by committees of supervisors, lawyers, political appointees, politicians, special use groups, and the general public before it was adopted. Apparently, a standard bolt of shingles is somewhat smaller than a log, but bigger than a small craft. Meaning that on some Washington streams, floating in small boats, inner tubes and kayaks may be prohibited.

Washington uses the federal test of navigability and also maintains that the state owns the beds of all navigable rivers. Under this standard, navigable waters include only streams capable of navigation for general commercial purposes, such as (you guessed it) floating shingle bolts. But navigable waters do not include "every small creek in which a fishing skiff or gunning canoe can be made to float at high water."[237] Washington courts have also determined as nonnavigable individual streams that had only been used by small boats for pleasure. And, the Washington Supreme Court has ruled that the quality of navigability of a watercourse need not be continuous, but the seasons of navigability need to be regular and of sufficient duration to serve commercial purposes.[238] On nonnavigable streams the public has no rights, and the owner of a nonnavigable stream can fence it off. In addition, Washington does not recognize portage rights across private land.

Wyoming

Wyoming allows the public to boat any nonnavigable streams that can be floated by any craft, even for pleasure, including a kayak. In the critical Wyoming court decision regarding the use of rivers for boating, Day v. Armstrong, the court stated, "The actual usability of the waters is alone the limit of the public's right to so employ them."[239] Additionally, the right to float the stream is accompanied by a companion right to portage around and over obstructions, including riffles, rapids, and dams. A limited right to access usable waters over private property may exist in Wyoming. All this makes Wyoming one of the friendlier states in the West for recreational water users.

A RIVER RUNS THROUGH IT

For many, fly fishing is a metaphor for life: if you can learn to fly fish, you've acquired the skills for living. In Norman Maclean's semiautobiographical novel (published in 1976), two brothers, one reserved and the other rebellious, are the sons of a stern Presbyterian minister. In his tale, fly fishing is the salve that cures the family's problems. Maclean (1902–1990) was raised in the Rocky Mountains and, as a young man, worked for the US Forest Service in logging camps around the region. Maclean attended graduate school at the University of Chicago studying English, gaining a doctoral degree in 1940. After World War II, Maclean taught at the University of Chicago and on retiring in 1976 wrote *A River Runs Through It*.[240]

In 1992, Maclean's novel was turned into a major motion picture starring Brad Pitt, Craig Sheffer, and Tom Skerritt. Robert Redford narrated and directed the film. *A River Runs Through It* became a breakout moment for Pitt, who starred as the rebellious son Paul Maclean. Redford, narrating, explained: "In our family, there was no clear line between religion and fly fishing." Set on Montana's Blackfoot River, the film uses the river as a backdrop to the tensions between the father and sons. Film critic Roger Ebert called Maclean's novel "one of the sacred books in the libraries of many people," and placed it alongside Walden and Huckleberry Finn as one of the masterworks of American literature.[241]

As a result of Maclean's novel and Redford's movie, fly fishing took off as an amenity sport where one might commune with nature and get away from it all. Anglers soon descended on many Montana and other western rivers. Sales of fishing gear exploded to the benefit of many local businesses. The novel became

Figure 2.163. In the acclaimed Norman Maclean novel, *A River Runs Through It*, fly fishing is a metaphor for life. Image courtesy of the University of Chicago Press.

something of an environmental parable and helped spur stream restoration and preservation efforts on the Blackfoot and other rivers. Without actually saying it, Maclean demonstrated that free-flowing rivers possess significant inherent value. For many, *A River Runs Through It* unveiled the possibility of interacting with sublime nature as they fished in the West.

RURAL ELECTRIFICATION

If it weren't for Franklin Delano Roosevelt and his New Deal, many rural Americans might still eat their dinner in darkness. When FDR became president in 1933, only about ten percent of America's rural population received electricity, while over ninety percent of their urban neighbors were connected to the electrical grid. As part of FDR's New Deal, in 1935 the administration pushed through legislation to create the Rural Electrification Administration (REA) over the objection of conservative interests who saw this program as socialist competition to private electrical utilities. Since private enterprise would not or could not serve electricity to rural areas, the Roosevelt Administration believed it was the duty of the government to provide social welfare to the countryside by seeing that rural residents had power. For farmers, the New Deal meant that they could have lights in their houses, refrigerate their produce, run motors, and pump water. With federal subsidies, power plants were built and power lines strung across vast regions of the American West. Hundreds of Rural Electric Cooperatives were organized as part of Roosevelt's efforts to help underserved communities. This revolutionized farming in America and led to the development of innovations such as the center pivot irrigation system.

Figure 2.164. In rural areas, like Colorado's San Luis Valley, irrigated agriculture on vast tracts would be all but impossible if it were not for rural electrification. Here both the groundwater well and the pivot are powered by electricity. Photo courtesy of the author.

RUSSIAN OLIVE

Since their widespread introduction by the Soil Conservation Service (the predecessor of the Natural Resource Conservation Service) to serve as windbreaks, Russian olives (*Elaeagnus angustifolia L.*) have naturalized in the western United States, invading pastures, riverbanks, and moist valley bottoms. It is particularly pernicious in wet saline environments. Here they have harmed forage production, reduced soil moisture, and have grown into dense thickets that shade out native species and alter riparian ecology. A native of southern Europe and Asia, this species was initially introduced to North America in colonial times and later in the West as early as 1903. Russian olives became favored as an ornamental species, and eventually many state and federal agencies promoted it for windbreaks, wildlife management, and erosion control, going so far as to give seedlings away to prospective users. By 1924, Russian olives began escaping cultivation in Utah. However, it was in the post–World War II era that the rapid spread of the species began. As information about the extensive environmental and agricultural impacts of Russian olive spread became apparent, the perspectives of land managers slowly changed, and its distribution and promotion gradually ceased. Eventually, as the spread became an infestation, in the early 2000s western States began placing Russian olives on their noxious weeds lists and requiring landowners to control the species.[242]

Figure 2.165. Russian olives that line the banks of the San Juan River near Bluff, UT, have grown so dense that they seem like an impenetrable barrier to the desert beyond. Photo courtesy of the author.

SALINITY

Salt is a necessary ingredient for life. But like all good things, too much salt is bad. As salt levels in fresh water increase, it starts to become saline, briny, or brackish. At a certain point, the salinity passes a threshold whereby harm begins accruing to plants and animals. Too much salt interferes with cellular processes. When this happens, growth is retarded, or if too much salt is present, the plant or animal consuming it dies. Everyone knows what the most common salt, sodium chloride—table salt—tastes like. But in nature, there are far more salts than our most common seasoning. Salt compounds form from the neutralization reaction between cations (positively charged ions) and anions (negative ions) so that the product is electrically neutral (no net electrical charge).

Except for distilled water, all water has some quantity of dissolved solids in it. Even water from the most pristine watersheds contains dissolved material. And many of these dissolved solids are various kinds of salts. To give an example, rainwater falling from the sky on the Continental Divide over Colorado's fourteen-thousand-foot peaks contains total dissolved solids in the range of 10 to 35 parts per million (abbreviated as either ppm or mg/l). As the water starts its way down the slopes of the Collegiate Peaks west of Buena Vista, it starts picking up dissolved solids. Here the total dissolved solids reach about 50 ppm, which is some

Figure 2.166. Salt crusts on the dry surface of the Desert Lake Reservoir near Cleveland, UT, suggest the power of evaporation for concentrating salt. Photo courtesy of the author.

of the purest water you can drink. But as the water descends the Arkansas River, it steadily increases in salinity as the water encounters the local bedrock and is used and reused by farms and cities. Once this water flows downhill through Pueblo, Colorado, and on to the Kansas state line, its salinity has increased to about 3500 ppm, a level that is inherently unsuitable for irrigation. In general, water with less than 200 ppm is considered excellent, 250–750 ppm as passable, 2000–3000 ppm doubtful, and over 3000 ppm entirely unsuitable for agriculture. Crop losses start occurring when the salinity reaches 750–800 ppm.[243]

Salts found in nature generally originate from the natural weathering of minerals in the soil or bedrock or come from fossil salt deposits in geologic strata that were once ancient seabeds. Salt concentrations also rise as water evaporates. Although there are too many salts to number, a few of the most commonly found salts in nature are worth mentioning and include sodium chloride, sodium sulphate (Glauber's salt), magnesium sulfate ("Epsom salt"), sodium bicarbonate (baking soda), calcium sulfate (gypsum), and calcium carbonate (lime or calcite).

SALMON

The identity of the American Northwest is inextricably tied to salmon and the salmon fishery. Archaeologists tell us that Indigenous people have been fishing for salmon for at least 7,500 years and probably much longer.[244] Their use of salmon is central to Indigenous peoples' beliefs and social institutions that developed in response to their interactions with salmon over these millennia. For these Indigenous people, salmon were remarkably abundant, and this abundance helps explain how a wild fishery, rather than domesticated animals, could support large sedentary native populations with complex social and political organizations that existed at the time of Lewis and Clark's expeditions. Salmon were undoubtedly the central keystone species that drove the Northwest's ecology and human history.

Salmon have a complicated natural history that determines how they propagate and are used. Salmon are anadromous, meaning merely that they spend part of their life cycle in freshwater rivers and lakes and another part in the salt water of the ocean and estuaries. Their fry hatch in mountain lakes and rivers and quickly migrate to the sea where they mature, later returning to the rivers where they were born to spawn again.

Throughout the centuries of Indigenous use, salmon populations seem to have remained remarkably stable.[245] All of this began to change in the wake of contact with European and American traders and explorers in the early nineteenth

Figure 2.167. Indigenous fishermen dip net fishing for salmon on the Columbia River at the Cul-de-Sac of Celilo Falls around 1957. Source: Corps-engineers-archives_celilo_falls_color.jpg Public domain image.

century. Once sizable populations of emigrants from the eastern United States and Canada began to settle in the Northwest and British Columbia, they began commercialized fishing on industrial scales that systematically undermined the previously sustainable salmon populations on northwestern rivers. After the turn of the twentieth century, when water developers saw northwestern rivers as potential sources for abundant hydroelectric energy, the situation for salmon quickly deteriorated. Both private and governmental actors actively began building dams, locks, and diversions that effectively blocked salmon migration routes across the region. With that, salmon stocks systematically degraded. As the salmon populations declined and eventually collapsed, the tribes, conservationists, recreational and commercial fishermen, and government agencies began decades-long maneuvering to stem the slide toward extinction. In particular, Indigenous people began fighting to recover salmon populations and reclaim the treaty rights to salmon. Leading the effort was a Tulalip tribal member, Billy Frank, whose work led to a major victory for the Indigenous peoples of the Columbia River Basin when a 1974 court ruling known as the Boldt Decision (after the presiding judge) recognized that the Tribes possessed treaty rights and were comanagers of salmon, along with state and federal governments.[246] Although all sides seem sincere in their desire to preserve the salmon, action toward their recovery remains challenging due to the complicated simultaneous ecological, social, political, economic, and legal action needed to make progress.

SALMON RIVER SWEEP BOAT

On Idaho's Salmon River, boatmen use a uniquely western vessel to navigate the powerful currents and rapids. The so-called "sweep boat" is named after the giant oars called "sweepers" that look a bit like oversized hockey sticks that boatmen affix to the boat's bow and stern. Boatmen use these oars to track through rapids. Salmon River sweep boats evolved from large rafts that river men once used to haul cargo on the Mississippi River. As prospecting and settlement came to the Salmon River and its tributaries, sweep boats were employed to carry supplies and equipment to mining camps and homesteads that were only accessible via the river. Early versions of sweep boats were outfitted with elevated platforms in the center to give the boatman a convenient perch to scout. Boatmen commonly lined their craft with green lumber to buttress the boat against the numerous submerged rocks that were an all-but-unavoidable hazard in the river.

Captain Harry Guleke (1862–1944) of Salmon, Idaho, built and piloted the first flat-bottomed sweep boat on the Salmon River. Guleke used his sweep boat to run the Main Salmon's rapids in 1896, making him the first known Euro-American to accomplish that feat. Guleke designed his boat to keep it running high in the water. He built it to be wide enough not to tip over sideways and long

Figure 2.168. Captain Harry Guleke and crew standing in a Salmon River Sweep Boat in 1919. Image courtesy of the Lemhi County Historical Society and Museum.

enough so that the vessel would not flip end over end when descending into a rapid.[247] Although a few companies still use sweep boats today, they mostly use them to ferry tourists down the river.

SALTON SEA

Spread out across the desert near Mecca, California, the *Salton Sea* supports fish and upwards of four hundred species of migratory birds in one of the largest hybrid ecosystems of the American Southwest. Although ancient lakebeds exist in this desert depression, they had long since evaporated by the time the first European visitors arrived. The contemporary lake, a human-made playa, was reborn as an outgrowth of the rush to reclaim the desert of California's Imperial Valley through irrigation. When high spring flooding breached poorly built dikes constructed by the California Development Company in 1905, the entire flow of the Colorado River poured unchecked for eighteen months into an ancient lakebed and established the modern-day Salton Sea. By the time engineers finally sealed the breach in 1907, the lake covered a 45 by 20-mile expanse of the desert. The lake soon supported fish and bird populations. Because of high evaporation rates in the desert, the lake has always had highly saline waters. However, in recent years, water diversions to Southern California cities and farms have reduced flows into the lake and increased the salinity to levels that now threaten wildlife. The receding lake has

Figure 2.169. A man looks out at the breach in the Colorado River as it flows into the Salton Sink to form the Salton Sea. Image courtesy of the Salton Sea History Museum. Public domain image. Photo source: http://saltonseamuseum.org/his_photos.html.

exposed the lakebed, which has created clouds of blowing toxic dust. This dust has resulted in increased incidences of asthma and cancer in nearby residents. The Imperial Irrigation District has implemented mitigation measures that include vegetation planting, irrigation, and surface roughening to control dust.[248]

SANITARY SEWERS

Many technological advances come with unintended consequences. By the 1870s cities were actively piping water to homes. Used water had to go somewhere, however, and initially homeowners and businesses simply drained it into their yards or nearby alleys. And once people began installing "water closets" in their homes the disposal problem rapidly became a nuisance. As more people installed toilets, groundwater tables rose to create soggy yards, or if the drainpipe led to an alley, the waste stream would become a health hazard and aromatic irritant. As cities grew and population density increased, these issues became acute. This led civil engineers to design *sanitary sewers*, or systems of underground pipes,

Figure 2.170. Stacked sewer pipes lie waiting for installation at the Kearney-Mesa defense housing project at San Diego in 1941. Photograph Russell Lee. Courtesy of the Library of Congress. Public domain image. Source: https://lccn.loc.gov/2017789528.

to collect the wastewater and remove it from the populated area and deliver it directly to streams. In the 1920s and 1930s, as the waste load continued to increase, the ability of streams to assimilate and disperse waste decreased, so cities began building wastewater treatment plants to clean up the effluent and improve stream quality. Of course, sewers are by no means unique to the West. As American cities developed, the need to remove human waste from occupied areas became an obvious necessity for hygiene and to improve the quality of life.

When cities installed their first sewer systems, they concurrently collected stormwater and runoff from streets. Once cities began building wastewater treatment plants the combined stormwater and sewer loads overwhelmed treatment plant capacity which led to untreated waste being dumped into nearby streams. That, and the added cost of treating street runoff, became an incentive to separate stormwater from other waste streams. Cities then began building parallel stormwater systems to limit water flowing down streets and direct those flows to creeks and rivers. Although sewer systems operate mainly outside of the public eye, these systems are one of the critical pieces of infrastructure that enable modern urban living.

SELENIUM

Go to the supermarket and you will find plenty of vitamin supplements containing *selenium* as one of the ingredients. Selenium, after all, is an essential trace element necessary for life that is naturally present in many foods. However, as the amount of selenium increases, the element rapidly becomes toxic. Excessive amounts of selenium include hair and nail loss, skin lesions, nausea, and nervous system anomalies, to name a few. But perhaps most troubling, too much selenium causes severe and often deadly birth defects in a wide range of mammals, including humans. In desert areas, particularly those with bedrock composed of shale and mudstone, the irrigation return flows may contain very high levels of selenium. When this selenium-laden water enters creeks or lakes, local wildlife populations may suffer. Most famously, in 1983, waterfowl in the Kesterson National Wildlife Refuge were discovered to be experiencing mortality, birth defects, and reproductive failure. These defects were eventually tied to elevated concentrations of selenium in irrigation return flows entering the refuge.[249] Selenium has also been suspected of harming native fish populations in the Colorado River where irrigation water runs off from agricultural districts. Because of selenium's severe effects on wildlife, it makes the element a good example of what economists call an "environmental externality." In other

Figure 2.171. Selenium poisoning has led to malformed waterfowl births at places such as the Kesterson National Wildlife Refuge where this element is concentrated in irrigation return flows. On the left is a malformed embryo from the Kesterson National Wildlife Refuge that is generally stunted and with no eyes and deformed bones in the right foot. On the right is a normal sibling embryo. Image courtesy of Joseph P. Skorupa of the US Fish and Wildlife Service. Public domain image.

words, selenium's release into tailwater is an unintentional or uncompensated side effect of one actor (the irrigator) that directly affects the welfare of another actor (fish and wildlife and the people who value them).

SIPHON

For anyone of a certain age that has owned a waterbed or a backyard pool, you will know that a *siphon* uses a bent tube or hose to move water from a higher container to a lower container, using gravity to keep the liquid flowing through the tube. The siphoning action begins when the tube is filled with water to build the hydrostatic pressure on the downstream end. This process, known as priming, creates a gravitational force on the downstream end of the pipe that allows the flow to start. When someone sucks on a hose to get water to begin flowing from a tank to a can, they are priming the tube so that it is full of the liquid.

Figure 2.172. An approximately twelve-foot diameter siphon on Denver Water's South Boulder Diversion Canal in 2006. Water flows through the pipe from the higher hill in the far distance, then it crosses the valley bottom, and moves up to the other side of the valley toward the viewer. Denver Water replaced and buried the pipe several years later. Photo courtesy of the author.

Once this happens, the liquid starts flowing through the primed pipe without any further assistance.

The Greeks and Romans knew about siphons and constructed them to help transport water to their cities. However, there were many practical problems in keeping their pipes sealed. Consequently, civilization had to wait for several millennia until piping and materials technology advanced far enough when, by the twentieth century, siphons become feasible for widespread use in modern water systems. At a small scale, farmers use curved metal tubes to move water up and over the banks of laterals and onto their fields. At the other end of the size spectrum, engineers construct giant concrete or steel siphons that transport vast quantities of water across valleys or under highways and rivers.

SMYTHE, WILLIAM ELLSWORTH

If the idea of reclamation ever reached the level of evangelical zeal, it was through the preaching of William Ellsworth Smythe (1861–1922). Born to wealthy parents in Worcester, Massachusetts, Smythe later migrated to Kearney, Nebraska,

Figure 2.173. William Ellsworth Smythe spent much of his life proselytizing for irrigated agriculture in the American West. This image is from William Ellsworth Smythe. *The Conquest of Arid America*. New York: Macmillan, 1911.

to work for a land development company. Once there, he became the editor of a local newspaper, where he traveled to New Mexico to see irrigation projects constructed. By 1889 Smythe was the editor of the *Omaha Bee*. When drought struck the Great Plains in 1889, Smythe saw farmers abandoning their land and shooting their livestock for lack of water. These events deeply affected Smythe, and he began penning a series of editorials about irrigation. He eventually became the chairman of the National Irrigation Congress. Smythe was in the right place at the right time to take up the mantle for promoting irrigation in the West. He traveled coast to coast on speaking tours promoting irrigation. His experiences speaking led him to founding the magazine *Irrigation Age* in 1891. The magazine became a vital source of information for farmers and promoters alike seeking information on opportunities and practices surrounding water development in the years before significant governmental support for irrigation became available. Along with that, he published articles in national magazines such as the *Atlantic* and *Harper's Weekly*. He published his first book, *The Conquest of Arid America*, in 1899, cementing his reputation as the most influential proponent for irrigation in the late nineteenth and early twentieth-century America.

Smythe ended up in San Diego, where he helped set up agricultural colonies in the Tijuana River Valley (now San Ysidro), near Glendale (Tujunga), and one focused on poultry called Runnymede. He returned to writing newspaper editorials and speaking for reclamation, irrigation, and dam building. He eventually moved to New York, where he became assistant secretary of interior for the veteran's land settlement at the end of World War I. He died in New York far from the lands he helped develop in the American West.[250]

SNOTEL

Newspapers like to print stories during April and May that might declare a high snowpack and anticipated flooding, or a low snowpack and potential drought. The information supporting these articles often comes from a US Natural Resources Conservation Service (NRCS) program called *SNOTEL* (for SNOwpack TELemetry). This automated system collects and reports snowpack information related to climate data in the western United States. The modern SNOTEL system stems from a congressional mandate dating back to the mid-1930s "to measure snowpack in the mountains of the West and forecast the water supply." Originally, NRCS staff collected field measurements of snow courses in western mountains.

Figure 2.174. A SNOTEL telemetry site managed by the Natural Resources Conservation Service. Photo courtesy of the author.

While they still collect field data, they use that information to verify the more extensive measurements collected at SNOTEL stations. The NRCS maintains multiple high-elevation snowpack measurement stations in western watersheds to provide various snow-related data. Individual sites operate unmanned and relay data remotely. Each SNOTEL site transmits snowpack data in near-real-time to agency staff. Over 730 SNOTEL stations operate in 11 western states, including Alaska. Typical SNOTEL sites contain pressure-sensing snow pillows, precipitation gauges, and thermometers. NRCS operates two master stations in Boise, Idaho, and Ogden, Utah, plus a central computing station in Portland, Oregon. SNOTEL data supports water resources, climate, and natural disaster (flood and drought) research. Hydrologists use the data in water supply forecasting and to support the resource management activities across the NRCS.[251]

SNOWMAKING

When visiting a ski resort in the West early in the season, you likely see cannon-like pipes blasting snow out onto the slopes. These machines are part of the

Figure 2.175. During the early ski season, snowmaking operations at resorts, like this one at Winter Park, CO, are continuous and take on the feel of an industrial operation. Photo courtesy of the author.

widely used snowmaking apparatus that resorts employ to bolster the snowpack. *Snowmaking* is a patented process that pumps pressurized, high-velocity streams of compressed air and water through a nozzle causing the water droplets to aerosolize and become at least partially crystallized as snow. This process works at air temperatures of 30° F or less. Resorts use this equipment to supplement snowfall, particularly during the early season, before significant amounts of natural snow accumulate. Although snowmaking appears simple and easy to perform, it is a costly and challenging process. The difficulties arise from the physical characteristics of water to supercool and remain liquid even when reduced to temperatures as low as -20° F. Still, resorts require significant infrastructure to move water up the slope to the point of snowmaking. Snowmaking alters the local hydrology by diverting water from streams or reservoirs and pumping it upslope to be frozen as snow. The reliance on snowmaking by ski resorts is likely to increase as climate change alters the timing and availability of snow during the ski season.

SOUTHWESTERN WILLOW FLYCATCHER

The southwestern willow flycatcher, *Empidonax traillii extimus*, lives in the southwestern United States, and like all sensible birds, winters in the warm climates of southern Mexico and Central America. It is rare throughout its range. Southwestern willow flycatchers frequent, as their name also implies, areas of dense riparian tree and shrub communities, including willow and other vegetation alongside rivers, streams, wetlands, lakes, and reservoirs. Southwestern willow flycatchers eat insects, seeds, and berries. It is a small bird, usually less than six inches in length, including its tail. They like to mate during late spring or early summer and typically build their nests in a vertical fork of a willow bush.

With its listing as an endangered species under the Endangered Species Act, federal law gives it the highest level of protection afforded to animals in the United States. By the time it landed on the endangered species list in 1995, the federal government had estimated that "well under a 1000 [nesting] pairs, more likely 500" nesting pairs remained. To help recover the species from the brink of extinction, the US Fish and Wildlife Service designated "critical habitat" in every state throughout its range. This designation affords a level of protection to its habitat and means that any federal action that might impact the bird requires careful evaluation and potential modification to ensure protection. If, for example, habitat exists—it's not necessary for the birds themselves to be present—at the site of a proposed dam, the project could get scuttled or

Figure 2.176. The southwestern willow flycatcher, a small bird with outsized political clout. Photograph Jim Rorabaugh courtesy of the USFWS. Public domain image.

significantly modified. One of the key factors cited for this bird's decline is the "general drying of riparian habitats." Biologists attribute the decrease to urbanization, water diversions, and dam construction. Other human factors include developments that eliminate or modify the natural processes "that establish and maintain these natural levels of dynamism, diversity, and heterogeneity in riparian ecosystems." These factors include livestock grazing, channelization, and land development. Conflicts over these impacts have created sharp tensions among conservationists, developers, and water interests. Another likely factor affecting the bird is the expansion of invasive tamarisk groves to the detriment of endemic willow species. But in a twist that has frustrated biologists, it turns out that at least some of these birds like to live among tamarisk and thereby complicate both tamarisk removal programs and species recovery efforts.[252]

To restore the southwestern willow flycatcher, biologists are taking a holistic approach toward ecosystem and watershed recovery. This approach means that a key focus for recovering the southwestern willow flycatcher is to restore or mimic naturally functioning riparian ecosystems throughout its range. Restoration efforts for species like the southwestern willow flycatcher demonstrate the power of laws like the Endangered Species Act to affect all manner of projects along rivers or within watersheds. It also suggests the interconnectedness of desert ecosystems and their fragility in the face of human development.

SPILLWAY

A critical structure in all reservoirs is its spillway. Engineers design spillways to convey water safely around a dam. Should floodwater enter a pool faster than the outlet works for the dam can drain it, the water level in the reservoir will increase. Eventually, the water will pass over the spillway in a controlled manner to protect the dam from damage. Engineers calculate the "probable maximum flood" for the watershed above any given reservoir when designing spillways. These runoff estimates allow engineers to place and size the spillway to minimize the likelihood of a catastrophic dam failure. Most of the time, spillways sit silently near the dam itself. But when they are needed, as happens now and then, these structures are the bulwarks against terrible flooding tragedies.

Figure 2.177. Here is the Oroville Dam spillway as project managers worked to prevent water from overtopping the dam itself. After nearly a decade of deepening drought where the reservoir volume was reduced to near record lows, the winter of 2016–2017 brought large quantities of rain and snow that resulted in a massive surge of water flowing over the spillway. Cavitation in the concrete caused the lower half of the structure to partially collapse. To control the situation, reservoir managers temporally activated a secondary spillway (at left) to reduce pressure on the primary structure. Image courtesy of the California Department of Water Resources.

Figure 2.178. Spillway of the Oroville Dam, CA, in happier times. Image courtesy of the California Department of Water Resources. Photograph by Paul Hames.

In late June of 1983, a record 120,000 cfs was flowing into Lake Powell from the Colorado River. Many dams have spillways set away from the main barrier, but huge dams, particularly dams with power plants, such as Glen Canyon, use tunnels bored around the sides of the structure through the bedrock abutments. Engineers avoid having water go over the top of dams, for the results could be disastrous. During this time, Glen Canyon's managers had the power plant and river outlet works running at full capacity. Even so, with the lake continuing to rise, their next step was to open the spillways. Outside of testing, this was the first time they operated Glen Canyon's spillways. On June 28, engineers increased the flow through the left spillway to 32,000 cfs. This quickly produced loud thumping noises, and the water coming from the spillway "turned an ominous amber color as the jet carried chunks of concrete and sandstone into the river." To stabilize the situation, they reduced the flow from the left spillway to 20,000 cfs and increased the flow in the right spillway to 27,000 cfs.[253] By July 2, Lake Powell was just 3.5 feet from the top of the spillway gates, and the water was rising three inches a day. If the water went too high, water would overtop the gates, and the officials would not be able to manage the spillway, and the entire dam might fail. The Bureau of Reclamation hastily installed four-foot-high plywood walls across the spillway gates. The unrelenting inflow continued to outpace the releases, and the water began overlapping the plywood. Workers quickly installed eight-foot steel plates. On July 8, officials had no choice but to increase the release to 92,000 cfs. Nervous engineers could hear thunderous rumbling from the spillways. For a few days, the situation was on the verge of spinning out of control. Fortunately, the runoff peaked and began subsiding, and the reservoir topped out on July 15, less than a foot below the top of the steel flashboards. This peak was only six feet below the crest of the dam itself. Hard work and a lot of luck narrowly avoided an unspeakable catastrophe.

Later that year, when inspectors could finally access the spillway tunnels, they discovered a gaping hole carved into the sandstone in the left spillway tunnel nearly 50 feet deep and 135 feet long. Further down the tunnel beyond the hole, they came across a boulder measuring ten by fifteen feet that fell out of the tunnel wall. Engineers concluded that during the release, the spillway tunnel had begun to "cavitate," meaning that the continued rhythmic surges of hydraulic energy had hammered and scoured the dam abutment. Had the hammering continued much longer than it did, Glen Canyon Dam itself might have collapsed, releasing some 27,000,000 acre-feet of water in a wall of water up to 700 feet high. Glen Canyon's spillways barely held, narrowly averting what would have been the worst human-made calamity in American history.[254]

SPIRAL JETTY

Robert Smithson's *Spiral Jetty* juts out like a rock tornado into Utah's Great Salt Lake. In building it, Smithson (1938–1973) took inspiration from spiral nebulas and the millions of stars they contain to construct this landscape-size conceptual art piece. A remarkable contemporary geoglyph, the *Spiral Jetty*'s monumental black basalt blocks penetrate the lake's reddish brine to create stunning visual contrasts. Visitors have described walking around it "a riveting experience" as they interact with the desert, lake, sky, and spiral. Smithson selected the site for this landscape sculpture at one of the few places on the Great Salt Lake where water comes right up to the mainland. Smithson felt "as if the mainland oscillated with waves and pulsations, and the lake remained rock still. The shore of the lake became the edge of the sun, a boiling curve, an explosion rising into a fiery prominence. Matter collapsing into the lake mirrored in the shape of a spiral. No sense wondering about classifications and categories, there were none."[255] In building the *Spiral Jetty*, Smithson used the landscape as a canvas. The *Spiral Jetty* lets visitors interact with the land, water, and sky in a stunning desert setting. Smithson began work on the *Spiral Jetty* in April 1972. *Spiral Jetty* remains a signature artistic statement from that era and perhaps the most loved example of land art.[256]

Figure 2.179. Here is Robert Smithson's *Spiral Jetty*, as seen from a satellite. On the right, three cars in a parking lot give a sense of the jetty's size. Image source: screen capture Google Earth.

SPRING

Sacred and tranquil. Pure and pristine. Naturally occurring electrolytes. Crafted by nature. Best-tasting. Nature's bounty. Healing waters. These are some of the slogans used to describe various springs across the western United States. Peel back the rhetoric, and you'll realize that *springs* are sites where groundwater reaches the land surface (or "daylights") in sufficient quantities to produce a stream of water. The local geological setting controls a spring's water quality. That and whatever industrial, municipal, and agricultural activities take place within the groundwater shed that supplies water to the spring further impact the quality. Spring discharge is closely related to the aquifer that supplies its water. In springs derived from local groundwater systems, a spring's flow is closely related to local precipitation and may fluctuate widely over the year. Also, these springs often have water temperatures that are near to the mean annual air temperature. In cases where the spring is receiving its water from regional aquifer systems, the flow rates are relatively stable. This stability arises because, in regional networks, the groundwater flow becomes averaged out over several years.

Springs form in a variety of geologic settings. A depression spring forms in a topographic low spot where the groundwater table reaches the surface. You may

Figure 2.180. Few springs are as dramatic as Vasey's Paradise, located deep within the Grand Canyon. Here, water gushes from the cliff face to feed the lush vegetation below. Photo courtesy of the author.

find contact springs where two geologic formations meet at the ground surface. In this instance, a low permeable formation such as shale underlies a formation that readily conducts water such as sandstone or gravel. A fault spring discharges water at a location that a geologic fault forces water flowing through an aquifer to the surface. Karst springs are typically associated with limestone formations where actual channels in the rock reach the surface. Sinkhole springs occur where part of a geologic formation erodes or dissolves away sufficiently to collapse into a hole where the groundwater table is exposed.[257]

Springs support a disproportionately high biological diversity relative to other landscape features. As such, they provide critical ecosystem services. Many small towns, farms, ranches, and isolated communities rely on spring water. People use springs across the West for spas and recreation. Springs have immense cultural, spiritual, and social significance with humans. Indigenous people revere springs and see them as living entities and the dwelling places for water deities. For Christians, springs represent renewal and life. Buddhists use springs as metaphors for rebirth. Muslims see springs within the gardens of paradise. Moreover, a spring is simply an excellent place to get cool refreshing water.[258]

STOCK POND

When American and European explorers first crossed the West's arid regions, they struggled with finding enough water for themselves and their livestock. Cattlemen particularly felt restricted by the constrained range for their stock due to limited access to creeks and rivers for water. Good forage remained unused because livestock could not wander too far from water sources. Ranchers responded to this scarcity by constructing small human-made lakes, or *stock ponds*, in seasonally dry gullies and drainages to collect rainwater and snowmelt for providing water to livestock. Construction of stock ponds became increasingly common during the early twentieth century. In the Dust Bowl era of the 1930s, ranchers desperate for water began building many more of these structures. Government subsidies, provided through federal laws such as the Taylor Grazing Act of 1934 signed by President Roosevelt, led to widespread construction of stock ponds.

Cattlemen usually build stock ponds in dry washes, arroyos, or isolated depressions where occasional rain or snowmelt causes intermittent flows. By damming these gulches, water can accumulate and provide a water source for livestock. Occasionally, ranchers used windmills to pump shallow groundwater into these ponds. Depending on the frequency with which they fill, stock ponds

Figure 2.181. Cattle grazing at a stock pond in a shallow wash located in Colfax County, New Mexico. Photo courtesy of the author.

may lie empty as mudflats, as moist weed patches, or produce wetlands where none had previously existed.[259]

STREAM NARROWING

Rivers and their floodplains are among the most dynamic geological systems on Earth. The channel patterns form over extended periods in response to numerous physical and biological variables. These include the amount of water and how it appears throughout a season, sediment type and supply, the size distribution of sediment (that is, the quantities of cobbles, sand, and silt present), stream gradient, bed elevation, surrounding bedrock characteristics, and local vegetation. In many stream systems, these variables change imperceptibly, in others dramatically, over time. Channels might remain stable for long intervals, but once one variable changes, the whole system can shift dramatically to a new state very quickly. One of the most dramatic consequences of hydrologic changes within a stream is *stream narrowing*, or the reduction in overall stream width over time.

Figure 2.182 and 2.183. Stream narrowing is an ongoing process occurring in many western streams and rivers. When viewed at any given time, stream narrowing is not particularly noticeable. However, by comparing photographs taken at the same location at different times, this process becomes striking. Take these two photographs of the San Juan River near Bluff, Utah, in 1925 and 2018. In the years between 1925 and now, the US Bureau of Reclamation built the Navajo Dam upstream, plus county and state governments built multiple bridges and roads. In 1925, the floodplain consisted of unvegetated, braided sand and gravel nearly three-quarters of a mile across. It has since narrowed to about 150 to 200 feet today, and exotic tamarisk and Russian olive along with cottonwoods, willows, and other underbrush dominate the remaining part of the floodplain. Even though the narrowing is dramatic, in the decades before 1925, imported stock had extensively overgrazed the bottomlands and harvested formerly extensive cottonwood groves, that no doubt contributed to the condition of the river at the time surveyors took the first photo. Upper photograph, courtesy of the USGS, W.T. Lee photographer, date 1925, Lee No. 3172 (public domain). Bottom image author.

Initially, when people began building dams in the West, they paid scant attention to the environmental effects of these projects. But once many projects were operational, the impacts started to pile up. Human activities, including dam construction, water diversions, the introduction of species such as tamarisk and Russian olive, and road and bridge building, are the primary cause for this narrowing. Among the biggest of these were the effects on stream channels downstream from dams. Once people began constructing dams, many of the variables that control the channel geometry changed drastically.[260] For western rivers, dams shift the discharge from infrequent high flow events to more or less constant flow throughout a season. Channels that were formerly scoured of vegetation gradually fill in with trees and underbrush so that in just a few decades, a narrow stream channel with a plant-filled floodplain emerges.

TAMARISK

Hike along a creek or river in the Southwest, and you often find yourself fighting through a nasty bramble of invasive tamarisk that makes progress all but impossible. In the Old World, tamarisk grows in arid and semiarid lands in a wide arc extending from Morocco to Sicily, across the Middle East, into South Asia, and up into China's desert regions. Introduced to the Southwest during the mid-1830s, tamarisk has naturalized and spread across the entire region. People initially brought tamarisk to America as an ornamental plant and to help control soil and stream bank erosion. We now consider tamarisk a noxious weed that forms monocultural forests along stream banks. Many people refer to tamarisk as saltcedar, or "tammies." They are deciduous trees and shrubs in the genus *Tamarix*. The genus *Tamarix* contains as many as fifty-four recognized species. The reason that tamarisk is also known as saltcedar is that the plant's deep root systems are very efficient in extracting groundwater. As the water transpires, salt, primarily sodium chloride, accumulates in the leafy vegetation. As leaves drop around the plant, the salt in the leaves modifies the soil chemistry in ways that benefit tamarisk to the exclusion of other plants. Tamarisk has invaded nearly every drainage system in arid and semiarid areas of the West, from California east to Texas and the Great Plains, and northwestern Mexico north to Montana. Water diversions, overgrazing, and off-road vehicle use contribute to the establishment of tamarisk. Across the West, tamarisk has stabilized stream banks so much that the sinuosity of rivers has measurably increased.[261] Another feature of tamarisk is its hardy thirst for water. Because tamarisk has spread so widely, and since it consumes so much water, water managers have

become increasingly concerned that this plant has singlehandedly reduced flows in western streams. As a result, heroic control efforts are now underway to manage this plant. Cutting, burning, and herbicide applications are standard control methods. Some biologists have taken the drastic step to introduce tamarisk beetles (*Diorhabda spp.*), an insect that weakens and kills tamarisk plants through continuous defoliation of its leaves. Fighting a biological invasion through the introduction of yet another exotic species creates significant angst among many biologists, as there is the genuine risk that this newest species will, in turn, become invasive or displace other native species.[262]

Figure 2.184. A riverbank reinforced by tamarisk roots. Green River, Desolation Canyon, UT. The tamarisk in this area has been weakened by the deliberate introduction of a beetle to control the plant. Photo courtesy of the author.

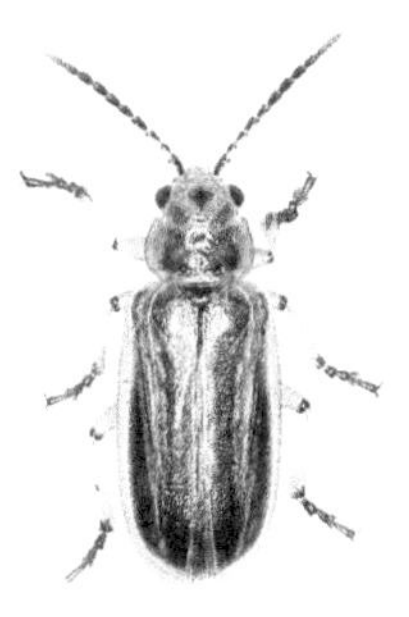

Figure 2.185. A tamarisk beetle at the US Department of Agriculture/Agricultural Research Service Grassland, Soil and Water Research Laboratory, Temple, Texas. Photo by James L. Tracy. Source Wikimedia Commons.

TERRACE

In the nineteenth century, geologists began working to understand stream geomorphology. They found that the flat, bench-like surfaces paralleling rivers resulted from erosional downcutting by streams themselves. These *terraces*, sometimes called "benches," are landforms common to river valleys.[263] Typically they are long and narrow, with the long side running nearly parallel to the river. Another feature of terraces is their mostly flat surface that slopes gently downward following the course of the river. Often several terraces parallel either side of the river like giant steps on the landscape.

Many terraces are subtle features, located only a few feet above the river level. Elsewhere, they form magnificent landforms, such as the Tonto Plateau and other benches within the Grand Canyon of the Colorado River. Along the Colorado River, some terraces are perched thousands of feet above the stream level and present a graphic testament to the millions of years of erosion needed to create its canyons. Although terraces are overwhelmingly erosional in origin, you can usually see a veneer of rounded alluvial gravel and cobbles, deposited when the former stream flowed at this higher level, when you walk across their surface.[264]

Figure 2.186. In this sketch, flat, gently sloping terraces formed by Left Hand Creek are visible. Colorado's Front Range rises in the background. The artist Henry Elliott drew *Terraces 5 Miles North of Boulder City. Looking South* in 1869, as part of the work he produced for the Hayden Expedition as it traveled along Colorado's Front Range. (Public domain document. Digital Archive: Previously Unpublished Sketches by Henry W. Elliott united with the Preliminary Field Report of The United States Geological Survey of Colorado and New Mexico, 1869, by F. V. Hayden. Edited by Kevin C. McKinney USGS Open-File Report 03-384)

In the American West, colonists quickly recognized that terraces were ideal sites for irrigation development. Their gentle slopes, proximity to water, and alluvial soils made agricultural development inevitable. By the mid-1860s, entrepreneurs were buying inexpensive land on terraces and digging ditches to bring water to these surfaces. Once watered, these entrepreneurs could sell productive farmland to new emigrants at a good profit. Within a few decades, vast areas of terrace lands were under irrigation and remain in productive use today.

TETON DAM DISASTER

On Saturday, June 5, 1976, workers arrived at Teton Dam to find a steady flow of water coming from near the toe, or base, of the structure. At first, the water was clear, which meant no sediment was moving within the dam. This leakage concerned the workers, but at this point, there was nothing to be alarmed about. Nonetheless, this was very unusual, and they quickly reported what they saw. At 9:00 a.m., the flow had increased to about two cfs, and the water had become slightly turbid. Various Bureau of Reclamation supervisors were alerted, and

Figure 2.187. This remarkable photograph was taken moments before the final breaching of Teton Dam on the morning of June 5, 1976. Jerry Durstellar, an employee of one of the dam construction companies, photographed the disaster from the first hints of trouble through the dam's final failure. Courtesy US Bureau of Reclamation. Public domain image.

a few began arriving on site. Just before 10:00 a.m., other leaks appeared, and these too were turbid. These leaks meant water was now flowing in pipes or channels within the structure—instead of filtering through the earthen dam as it should have been. Disturbingly, workers could now hear the water moving within the structure. The situation was swiftly becoming critical. By 10:30 a.m., the flow had increased, and the leak was now eroding a hole upward along the dam face. A few workers were heroically, but futilely, trying to dump rip rap into the hole to stem the flow. The water volume in the leak rapidly increased. Within minutes, a crack began developing on the dam face, and then two M-K bulldozers fell into what was now a gaping 100-foot-wide hole. The machine operators luckily escaped. Then, at about this time, workers began hearing loud bursts of noises, and a whirlpool opened up in the reservoir on the upstream side of the dam. As the clock approached 11:00 a.m., the vortex rapidly grew. At the dam site, people frantically called authorities downstream to alert them about the imminent dam collapse. In a matter of minutes, the whirlpool doubled, then tripled in size. At this point, the water was agitated, frothing, and muddy. Suddenly, a large part of the dam—20 feet wide by 20 feet high—sluffed off into the suckhole in one giant chunk. This failure caused the whirlpool to boil even more violently. Within a minute or so, the top section of the dam, undermined from below, suddenly dropped. With that, the structure catastrophically collapsed into the cascading water. By about 11:15 a.m., Teton Dam was gone.

A wall of water containing nearly the entire contents of the reservoir—more than 234,000 acre-feet of water flowing at over 2,000,000 cubic feet per second—hurtled downriver, causing massive loss of life and destruction. After the breach, it took about 5 hours to drain the formerly full reservoir. When the sheriff of Madison County, in Rexburg, Idaho, received the emergency call at about 10:50 a.m. that morning, he did not immediately understand the severity of the warning, but hedging his bets, he began telephoning people who lived in the potential flood path. As the water rushed downstream, the channel filled to a depth of at least 30 feet for as long as the eye could see. The inundation downstream was disastrous, with about fourteen persons killed. Wilford, Sugar City, Salem, Hibbard, and Rexburg, Idaho, all received severe flooding. Authorities estimated property damage of up to $2 billion. Nearly 13,000 cattle perished. That plus the failed dam itself cost about $100 million to build. The federal government eventually paid over $300 million in claims.

Teton Dam and reservoir were the central features of the Teton Basin Project, a multipurpose program by the US Bureau of Reclamation embodying flood control, power generation, and supplemental irrigation water supply. Site work on the dam began in February 1972. Before collapsing, it was a 305-foot-high earth-fill dam. At the time of the disaster, the Bureau of Reclamation was filling the dam for the first time. A commission examining the events attributed the tragedy to a combination of geological factors and design decisions that, taken together, permitted the failure to develop. Many point to this calamity as the end of the big dam era in the United States. Since then, no one has mustered the political will to rebuild the Teton Dam.[265]

THERMAL SPRINGS

Hydrologists know of nearly 950 thermal springs scattered across the eleven western states. Over the years, entrepreneurs have converted many of these into commercial hot spring resorts. Commonly referred to as warm or hot springs depending on their relative temperatures, *thermal springs* discharge water at temperatures above the mean annual air temperature for that locality. These temperatures mean that the heat warming these springs comes not from the atmosphere but sources deep in the Earth. Thermal springs usually contain highly mineralized water. As heated water travels underground, it picks up minerals from the surrounding rock. When this water reaches the surface, the

Figure 2.188. William Henry Jackson's 1871 photograph of Mammoth Hot Springs on the Gardner River in what is now Yellowstone National Park. Image courtesy of the National Park Service. Public domain image.

acidity of the water changes, and mineral deposits form. Thermal springs almost always have a relationship to volcanic rocks that cause the nearby ground and water to heat up. Many thermal springs have a very consistent temperature, flow, and mineral content. With sufficient temperatures and discharge, hot springs support commercial resorts. When huge amounts of heated water are present, it becomes possible to develop geothermal power plants.[266]

TIE DRIVES

One often thinks of the North Country when talking about logging and the great tie drives of the nineteenth century. And while lumbermen did decimate the northern forests, we often overlook similar takings in the West. Beginning with the 1849 gold rush, Argonauts exploited forest resources for everything from mines, flumes, home and fence construction, charcoal production, heating, and for whatever else they needed. But forest exploitation got into high gear as railroad men constructed the transcontinental railroad and all the other lines that followed. They needed railroad ties, vast numbers of ties, to put down all that track. To get those ties, they clear-cut primordial forests with utter disregard for the ecological consequences. And to transport them from the forest to the railheads, they often drove them down the western rivers by the tens of thousands

Figure 2.189. Railroad tie drives, such as this one on Douglas Creek, Wyoming (early 1900s?), caused significant damage to riparian corridors. Image courtesy of the University of Wyoming American Heritage Center, Medicine Bow National Forest (Wyo.) records, Accession #03654, Box 17, Negative Number 18763.

at a time. By the late nineteenth century, once vast western forests lay denuded of all usable trees. Tie drives scrapped river channels clear of vegetation. Some channels remain scoured to this day. Although secondary growth has filled in clear cuts, they seldom resemble the primary old-growth forests. Few practices exceed the destruction wrought upon western rivers by the nineteenth-century tie drives.[267]

TINAJA

In the southwestern deserts, ephemeral water pockets or pools often form at the base of waterfalls, in wind-scoured bedrock, or other naturally occurring basins in bedrock. These tinajas can develop wherever water from creeks, springs, seeps, snow, or thunderstorms collects. Often the best-developed tinajas are found in bedrock canyons where high-velocity water flowing through narrow channels has scoured a pool. The term comes from the Spanish *tinaja* (via the Moors of North Africa), which refers to porous water jugs or tanks that chill water through evaporation. Because of its Spanish origin, you will find the term in use in areas that were formerly parts of Mexico, such as Arizona, the southern California deserts, and New Mexico. In areas where there was little

Figure 2.190. Here is a tinaja that formed in a bedrock canyon in Utah's San Rafael Swell. Photo courtesy of the author.

Spanish influence, the term "water pocket" is often used for the same feature. In extremely arid regions, such as in the Sonoran Desert, tinajas provide essential water to wildlife.

Additionally, some fascinating organisms inhabit these water bodies. Anhydrobiotes, organisms that tolerate virtually complete dehydration, survive in these desert waterholes. These include various fungi, algae, and perhaps most famously the "fairy shrimp," those tiny freshwater crustaceans that are related to lobsters, shrimps, and crabs.

And, in the time before humans could drive long distances in their cars, or ride in the comfort of a train, these desert water sources provided essential moisture that made survival in otherwise barren landscapes possible. As a result, the knowledge of where to locate these tinajas often made the difference in life and death to desert travelers.[268]

TRANS MOUNTAIN DIVERSIONS

As the name implies, a *trans mountain diversion* moves water from one river drainage to another. There are many types of interbasin water transfers. Still, the sheer size, expense, technical sophistication, and ecological and economic ramifications of trans mountain diversions put these projects in a class all their own. Early projects began in Colorado by constructing ditches in the high country. Eventually, the ability to tap remote watersheds led water developers to bore tunnels through the mountains. By building tunnels, engineers could reduce the overall distance needed to transport water (by avoiding circuitous open ditch routes through the mountains). Furthermore, because engineers could capture water lower down in a basin, they could capture more water than otherwise available higher up.

Trans mountain diversions are often extremely controversial in the basin from which the water originates. For example, dewatering the Upper Colorado River has reduced stream flows needed for fish and wildlife, rafting and recreation, and for the cities, towns, and farms whose growth is limited by removing water from their area. Legislators from the affected regions have introduced so-called "basin of origin" legislation to restrict such diversions or offset their impacts. But, municipalities that use this water often stymie any legislation to curtail trans mountain diversions. Trans mountain diversions are often lightning rods that expose urban-rural conflicts over environmental, economic, and social concerns.[269]

Figure 2.191. The Northern Colorado Water Conservancy District built the Alva Adams Tunnel to divert water from the Colorado River and transport it beneath Rocky Mountain National Park to the South Platte River drainage. Shown here are workers ending a shift during the construction of the tunnel's east bore. Image courtesy of the Northern Colorado Water Conservancy District.

TUNNELS

Among the many structures to convey water, tunnels have a unique role. Long-distance tunnels have allowed water diversions to move water under entire mountain ranges. Among the best known are the various tunnels that divert Colorado River water under the Continental Divide to the South Platte River. The Roberts Tunnel, a facility of Denver Water, for example, is approximately 28 miles long and is a major component of Denver's water supply. The Colorado–Big Thompson Project's Alva B. Adams Tunnel is the key component of Colorado's largest trans mountain water project. At 13.1 miles long, it delivers Colorado River water under Rocky Mountain National Park to the Front Range of Colorado. Elsewhere, the Gunnison Tunnel, christened in 1909, opened up large tracts of land in western Colorado to irrigation development.

Generally, these water tunnels are intricately related to trans mountain diversions. Their construction is often the lightning rod that creates controversy between the basin of origin and the basin to which the water is delivered. Issues

Figure 2.192. Completed in 1909, the Gunnison Tunnel was one of the first three projects authorized under the Newlands Act. A primary feature of the Bureau of Reclamation's Uncompahgre Irrigation Project, the tunnel brought water from the Gunnison River to open up lands in the Uncompahgre River Basin to irrigation. This project was so important for western Colorado that President William Howard Taft presided over the dedication ceremonies. Image courtesy of the US Bureau of Reclamation. Public domain image.

surrounding water tunnel projects often mimic the environmental debates of the big-dam era. Accordingly, water tunnel construction usually involves protracted licensing processes that pit environmentalists and water users from the basin of origin against the water developers proposing the project. In recent years, water developers have built few water tunnels in the western United States.

TURF GRASS

It seems that for many people seeking the American Dream in the West, the aspiration is to own a home in the suburbs surrounded by a patch of deep green grass to call your own. And so those of us who live in the West see that upwards of 50 percent or more of local water supplies go to irrigating and maintaining turf grass, urban lawns, and golf courses. And with these lawns, we have whole

Figure 2.193. A sod farm in Weld County, CO. Photo courtesy of the author.

armies of lawn maintenance companies, lawnmower salesmen, hardware box stores, and chemical fertilizer and pesticide companies serving local homeowners to keep their grass green. Plus, untold numbers of loved ones nagging their spouses to mow their lawn.[270]

UNION COLONY

On the parched steppes of eastern Colorado, utopian idealists launched Colorado's Union Colony, one of the great experiments in nineteenth-century agrarian development. Here, migrants from the eastern United States succeeded in building a dynamic society by focusing on communal action to build ditches, roads, farms, and the city of Greeley, Colorado. Because of its close connection to its primary benefactor, Horace Greeley ("Go west young man!"), the colony's main town bears his name. Greeley was an active supporter of European Socialism and at one time even employed both Karl Marx and Frederick Engels as journalists for his newspapers.[271] Another idealistic follower of Greeley, Nathan Meeker, became the central organizing force for the colony. The Union Colony succeeded because the colonists used collectivized labor to build several large ditches that took water from the Cache la Poudre River and delivered it to their new farms. Ironically, although the colony itself espoused socialist principles

Figure 2.194. Union Colony colonists employed collectivized labor and oxen teams to dig the 13-mile-long Greeley No. 3 Ditch. In addition to oxen, men used pickaxes, shovels, and plows to excavate the ditch. Because colonists located the Union Colony on a dry terrace, completion of the canal was essential to supply water for domestic and agricultural purposes. Work on the ditch began in May 1870, and workers completed it in about a month. The Union Colony organized its labor along communal lines to construct the ditch, achieving a feat that was otherwise unobtainable by individual labor. Public domain image, source: http://www.greeleyhistory.org/pages/number_three_ditch.html.

in its creation, colonists turned to some of Colorado's early capitalist oligarchs to acquire their land. In particular, former governor John Evans and associates acquired railroad land that they had helped appropriate from the region's Indigenous communities using military force. Colonists bought the railroad land and began their settlement where Indigenous people had grazed their ponies scarcely ten years earlier.[272]

US ARMY CORPS OF ENGINEERS

US Army Corps of Engineers, "Army Corps," or commonly "Corps," can trace its history back to George Washington and the American Revolution when the general appointed the first engineer officers of the Army on June 16, 1775. Then the Jefferson Administration organized the Corps in 1802 as a separate branch of the Army. Initially, the Corps built defensive works, but its mandate quickly expanded to include the construction of coastal fortifications, roads and canal

Figure 2.195. The US Corps of Engineers built the Bonneville Dam on the Columbia River between 1934 and 1937. The project gets its name from Army Captain Benjamin Bonneville, an early explorer and fur trapper who helped blaze the Oregon Trail. Bonneville Dam, like many others built by the Corps in the 1930s, was a product of the New Deal. Shown here is a detail of the massive spillway structure on the dam. Source: Downstream face of dam spillway, Bonneville Dam, Columbia River, Library of Congress. Image HAER ORE 26 BONV 2 F 3, public domain.

surveying, mapping, and removal of navigational hazards, among other things. Throughout the nineteenth century, the Corps' portfolio expanded and acquired significant responsibilities over civilian infrastructure. By the twentieth century, the Corps became the lead federal flood control agency. It also moved into the construction of dams for hydroelectricity, levees for rivers, and navigational locks. Today, it is even deeply involved in numerous environmental matters by its permitting of dredge and fill activities (the so-called "404 permit"), which affect virtually all of America's wetlands and waterways.[273]

In the West, only the Corps rivals the Bureau of Reclamation in the number and scope of water projects that it operates. Corps has major dams that straddle state lines, such as the John Day, The Dalles, and Bonneville on the Columbia River between Washington and Oregon. Additionally, the Corps operates seventeen dams and reservoirs in California's Central Valley, the Alamo Dam in Arizona, Chatfield Dam in Colorado, and Fort Peck Dam in Montana, to name but a few. Between the projects it operates and the regulatory oversight it possesses, the Corps often finds itself at the center of complex, controversial, and politically fraught water policy and infrastructure matters in every western state.

VIRTUAL WATER

Some of the most essential supplies of water circulating in the western United States are invisible. You will not see this water flowing down the region's rivers, filling its aquifers, or standing in its lakes. But this water helps many more people live in the West than otherwise possible without it. Specifically, I refer to the water used to grow commodities such as grain. Likewise, this water helps manufacture industrial products including microchips and other goods. James A. Allen, the geographer who first recognized the importance of this esoteric water category, calls it *virtual water*.[274]

The notion of virtual water has profound significance for arid regions, including the western United States. Consider this: to grow one cubic foot of grain, you need about 1000 cubic feet of water.[275] Allen saw the nexus between food security and water. He pointed out that people living in an arid climate could get away with living there if they imported considerable quantities of food from water-rich regions. They might also import manufactured products that require lots of water in their making. By importing commodities and products that require considerable quantities of water to grow or make, an arid region experiencing a water deficit or drought might get by or even thrive. These commodity imports mean that some other area's water gets used far from the arid region that consumes them. Virtual water allows an arid region to survive on lesser quantities of water so long as food and other necessities come from elsewhere. Allen also pointed out that virtual water helps political leaders avoid confronting water deficits that would otherwise become existential problems. Take the intermountain West, a net importer of food.[276] States such as Nevada or Arizona don't need to use all their scarce water for agriculture because they can import wheat from water-rich states like Kansas, corn from Iowa, apples from Washington, rice from Arkansas, and so on. Politicians in Nevada or Arizona thereby experience far less pressure than otherwise when dealing with rural to urban water transfers. Sure, rural to urban water transfers are big issues in those states, but can you imagine the controversy if it meant that some people might not eat? In some regions of the world, this effect is so profound that virtual water has relieved the pressure on nations' water resources to the extent that they have avoided war.[277]

Figure 2.196. An arid region avoids needing to use its scarce water supplies for various crops and manufacturing by importing commodities from water-rich areas. Geographers and economists call the water embedded in these commodities virtual water. This invisible imported water allows people to live in arid regions that would otherwise have insufficient water to grow all the food they need locally. Here, the cornrows in Buchanan County, IA, extending as far as the eye can see, support people far from this water-rich state. Photographer Carol M. Highsmith, 08-16-2016. Cropped from original. Public domain image courtesy of the Library of Congress. Image https://www.loc.gov/item/2016630438/.

Another political ramification for virtual water is that it creates a disconnect between human communities and the water resources they use.[278] Ecological damage, soil erosion, pollution, and adverse social impacts can all occur far from consumers and make them less aware of the consequences of their purchases.[279] Cars, consumer electronics, and clothes are some of the products that use water in their production, often on continents outside of North America. Moreover, other resources, such as electricity, frequently require significant quantities of water to produce and may occur in an area far removed from the point of consumption.[280]

The internet was still in its infancy when James A. Allen coined the term virtual water. I am not sure how much he thought about the vast networks and interconnections that the internet represents. But in hindsight, virtual water is a pretty good metaphor for modern societies' interconnectivity and the use of critical water supplies far from where the consumers reside. It seems that there is a lot of virtual water flowing around for something that you can't see, but it helps explain why western cities seem to be living beyond their (water) means.

WATER BUFFALO

> "Once recognized as the fiercest beasts roaming the wild open wetlands of Asia, water buffalos earned their reputation as aggressive warriors able to travel long distances and engage in dangerous stampedes. Still stampeding, the modern Water Buffalo . . . [is] still hunting for water, but these days the search for supply is typically sought through legislation and state regulations."—Jennifer Findley, BC Water News[281]

In the American West, water buffalo is the derisive term given to water developers, politicians, and technocrats. The name apparently comes from the visual image of the Asian water buffalo wallowing in muddy waters. It suggests that water developers are single-mindedly preoccupied with the imperative to build new dams and reservoirs. An excellent example of the term comes from George Sibley who wrote, "All this began to come home to the Bureau, and the

Figure 2.197. Here is a quintessential gathering of twentieth-century water buffalos. Colorado Gov. Clarence J. Morley, seated at left, signs legislation approving the Colorado River Compact and the South Platte River Compact in 1925. Delphus ("Delph") Carpenter, Colorado's chief negotiator for both compacts, stands front center. Carpenter originated the idea that states appropriate the waters of shared rivers through negotiated compacts instead of conducting ongoing legal battles. Courtesy Denver Public Library, Special Collections.

rest of the brotherhood of the water buffalo when in 1968 they had to give up two big Grand Canyon power dams to get the Central Arizona Project through Congress."[282]

But not all water users consider being called a water buffalo a pejorative term. Some people at the Central Arizona Project have asked: "Just what is a water buffalo? In Arizona, they are those iconic figures who had the foresight to plan ahead to meet the water needs of a growing desert community. Without them, we might not have the Central Arizona Project and the state's water situation might be bleak."[283]

WATER CONSERVATION

All sorts of technologies and strategies fall under the umbrella of water conservation. The basic premise is to reduce the amount of water used in any given practice so that the saved water is available for use elsewhere. The general public never sees many water conservation practices: the lining of canals, reducing leaks in pipelines, and the installation of drip irrigation in agriculture, to name

Figure 2.198. A xeriscape garden in Cave Creek, AZ, shows how designing landscape features with the local climate and environment in mind can reduce outdoor water demands. Photo courtesy of the author.

a few. Other methods are very well known and include the installation of low flow toilets and showerheads along with replacing yard grasses that consume large amounts of water (such as bluegrass) with drought-tolerant landscaping. Municipal water conservation also has the related benefit of saving considerable energy because the utility pumps less water.

One of the ironies of water conservation is that efficiency measures often have unintended consequences that can cause harm to other users or the environment. When a farmer, city, or industry installs devices to improve efficiency, they almost always want to use the saved water to extend yields, supply it to new customers, or expand output. You can't blame them for that, but this means that the overall consumption might increase, leaving less water for other users or the environment. Likewise, when a western city initiates programs to curtail wasteful irrigation of bluegrass and has its citizens install buffalo grass or xeriscapes, the water provider invariably shifts the saved water to new taps that serve indoor uses. As a result, more people rely on the existing water infrastructure that supplies the city. Although this is generally good, it also means that in a drought, there is more vulnerability in systems that contain less slack in the form of expendable bluegrass turf.[284]

WATER FEATURES

Go down to the street in front of the Bellagio Hotel in Las Vegas any evening and you can join the tourists waiting for the extravagant water show to begin. The spraying water and the associated light show tease people into the casino for an evening of fun. Here at the Bellagio, you might enjoy the largest and most famous water feature operating in the American West.

"Any artificially constructed display of water—either active, such as a fountain or a waterfall, or passive, such as a pool or canal—is a water feature in developer slang," explains Dolores Hayden in her guide to sprawl. Developers often add lakes, fountains, or other displays of water in arid areas to soften harsh and xeric landscapes and make their projects appear more visually appealing to potential customers who are more accustomed to wetter climates. From marinas along the shores of Lake Powell, past the fountains of Las Vegas, to subdivisions lining former aggregate mines, water features have become a mainstay of the altered western landscape.[285]

Municipal codes sometimes limit the sources of water used in active displays to reclaimed, grey, or reused water. Still, water features create an illusion of abundance and wealth, even in water-scarce areas.

Figure 2.199. At the Bellagio Hotel and Casino fountains in Las Vegas, NV, water leaps into the air in one of the largest water features in the American West. Although the water in the Bellagio fountain is recycled, the opulence of seeing so much water sprayed dramatically into the desert sky obscures the necessity for conserving water in this arid state. Photo courtesy of the author.

WATER GLYPH, OR CUP AND CHANNEL PETROGLYPHS

Of the many enigmatic petroglyphs scattered about the Western deserts, the "cup and channel" petroglyphs, sometimes referred to as "water glyphs," are among the most mysterious. We mostly find these cup and channel petroglyphs in remote locations along the Arizona Strip region of northern Arizona and southern Utah. Although several variations of these petroglyphs exist, their general style is that of a circle bisected by a line. But double or partial circles and dots within the circle are common as well. Sometimes the bisecting line is curved instead of straight. Indigenous people carved these glyphs deeply, much more so than virtually all other petroglyphs. Most water glyphs extend three to six feet across, but archaeologists have found a number of "mini" water glyphs that are fewer than six inches in length. Observers find these petroglyphs located near cracks, crevices, natural basins, or along the edge of a rock ledge. Archaeologists know of at least 370 of these petroglyphs. Just why Indigenous people carved these petroglyphs remains an enigma.

Figure 2.200. This typical cup and channel petroglyph located in northern Arizona might be an artifact from an Ancestral Puebloan water ritual. Photo courtesy of the author.

Hypotheses for the function of the cup and channel petroglyphs include prehistoric navigational markers to water sources, solstice markers, and ceremonial water channels. It is the water source hypothesis that gives rise to the "water glyph" term that some rock art researchers attribute to these petroglyphs. Some who espouse the water glyph hypothesis believe the line bisecting the circle points to springs, tinajas, or other water sources.

However, each hypothesis has significant problems associated with them. For example, the idea that these point to distant water sources is, at best, suspect. After all, why would an Indigenous person peck a petroglyph on a remote mesa top far from regular travel routes as a marker toward water? And if the Indigenous person creating the petroglyph were local to the area, his band would already know where all of the local water sources are and how to get there, so he would not need specific pointers. And it is doubtful that someone would make markers for foreign travelers through the area, especially on remote mesas. Moreover, only a few of these markers appear to point toward water, and even these seem coincidental. Research on the solstice marker hypothesis suggests some of these petroglyphs do line up with the sunrise on the winter solstice. However, many others do not.

It seems that archaeological and ethnographic evidence best supports the ceremonial water channel hypothesis. Evidence for this includes channel depths and slope sufficient to move water, patterns at the glyph sites to suggest a channeling function (moving water over a ledge to a specific spot). A large number of these sites have nearby habitations and their proximity suggests that the petroglyphs may have had ceremonial uses for the inhabitants, specifically as water shrines. Besides, historic Pueblo accounts confirm the importance of water in their ceremonies. But even this evidence is inconclusive, as it assumes that the various cup and channel petroglyphs have had consistent uses over time and space. It seems clear, however, that these petroglyphs had ceremonial significance and may have had dual roles in water ceremonies linked to solstice events. Unfortunately, like many things in the archaeological record, the real purpose of these enigmatic petroglyphs will likely remain shrouded from our full understanding.[286]

WATER GRAB

Growing cities are known for their aggressive tactics to secure water supplies for development. We term these strategies *water grabs* when cities employ secrecy, questionable ethics, aggressive politics, or unscrupulous water brokers to obtain water rights or shares from ditches, farmers, or distant watersheds. Perhaps the most famous water grab was William Mulholland's successful effort to secure water for Los Angeles from California's Owens Valley. Such practices are not relics of the distant past. In 1986, the City of Thornton, Colorado, made bold moves to "buy and dry" about 120 farms covering nearly 7,000 acres in Weld County so that it might transfer the water to municipal uses. Around the same time, the city of Aurora, Colorado, purchased water rights to dozens of farms near the town of Rocky Ford along the Arkansas River, leaving dried up farmland as its legacy.

More recently, the Southern Nevada Water Authority has proposed drilling wells and building a pipeline system to convey groundwater from the Spring, Cave, Dry Lake, and Delamar Valleys of central and eastern Nevada to southern Nevada. Residents of the targeted valleys worry about the impact on farms and ranches, local environment, and regional economy and fret that water leaving their counties will fuel growth far away without any local benefit. For decades, a consortium of ranchers, rural county commissioners, environmentalists, Indigenous leaders, and others opposed the pipeline. These groups coordinated their opposition with organized protests and legal actions. Then, in early 2020, a Nevada District Court

Figure 2.201. Participants of a parade during the Snake Valley Festival protest against the Southern Nevada Water Authority's proposed water pipeline. If built, the pipeline would have allowed groundwater development in rural Nevada for development in the Las Vegas area. After facing years of resistance, the Southern Nevada Water Authority scrapped the proposal. Image courtesy of Gretchen Baker, Snake Valley Festival.

ruled against major aspects of the plan, and the Southern Nevada Water Authority abandoned its decades-long effort to build the pipeline.[287]

WATER PURIFICATION: WATER AND WASTEWATER TREATMENT PLANTS

In ancient times, people strove to acquire water from clean sources. They knew that some water sources were healthy and others were not. The Romans, for example, built the famous aqueducts so that they might get clean water from remote sources to their cities. But for millennia, the reasons why some waters were good and others bad remained a mystery. Although by the 1700s a few cities in the United Kingdom filtered their water, they only did this to remove visible debris. Everything changed in 1854 when the physician John Snow identified the

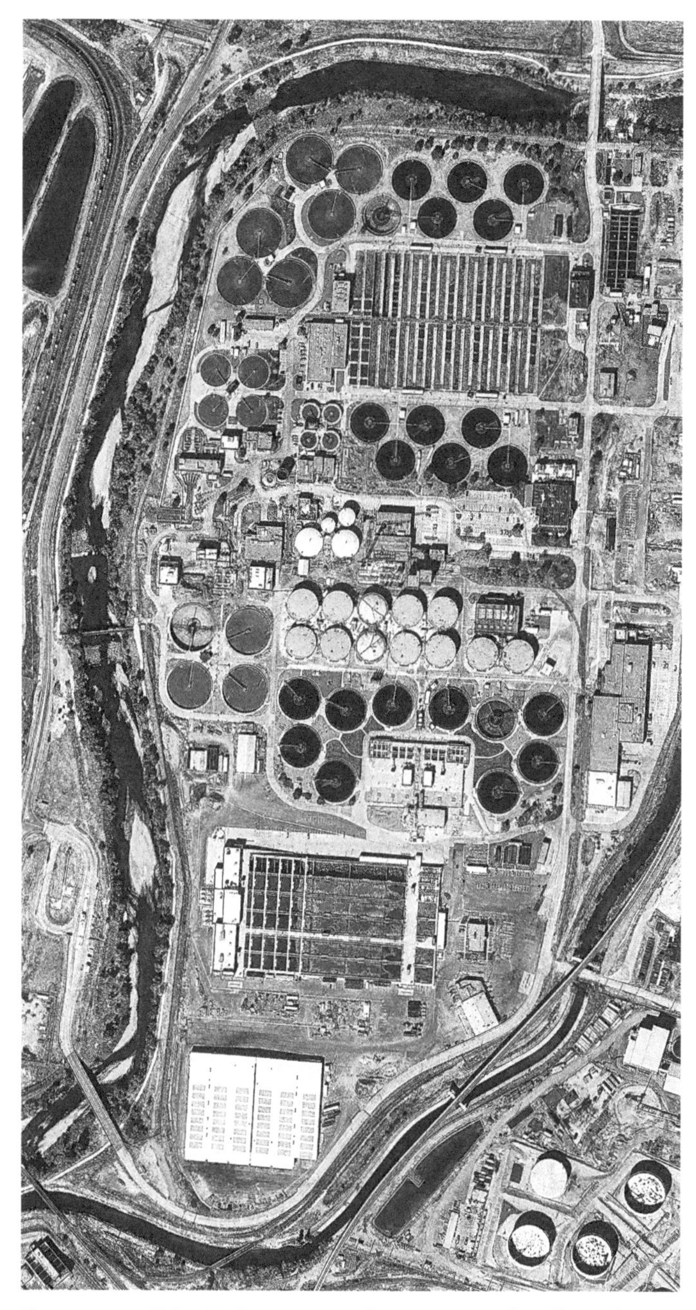

Figure 2.202. Urbanization can transform water quality. Shown here is Denver's Metro Wastewater Treatment Plant. At this plant—especially during winter months—sewer effluent outflows often exceed the entire flow of the South Platte River. Image source: screen capture Google Earth.

spread of cholera via contaminated drinking water from a well in London. Snow's work allowed scientists to grasp that invisible organisms living in water might harm or kill. Louis Pasteur then pioneered the germ theory of disease in the early 1860s. Pasteur's work helped form the foundation for modern systems to purify water for human consumption. It was from these advances that our contemporary water treatment plants emerged. We now design modern cities with a water treatment plant that purifies and distributes potable water to inhabitants and a waste collection system that ends at a wastewater treatment plant to sanitize the water before releasing it back to rivers, lakes, or the ocean.[288]

Modern water treatment plants vary dramatically in size and capacity and range from small operations supplying water to just a few homes to giant industrial-sized plants serving water to large cities. But their fundamental components remain relatively consistent. In general, when raw (untreated) water enters a plant it first moves through a variety of progressively finer filters (sand, carbon, and then micron filters) to remove debris, suspended solids, and even dissolved organic matter. High pressure pumps then push the water through membranes to remove most of the remaining material in the water. The plants then use strong ultraviolet lights to kill any remaining microorganisms. Finally, workers inject small amounts of chlorine into the water (in Europe, they commonly use ozone) to further ensure that no remaining microbes linger. After this, the water enters the pipes that deliver it to our homes.[289]

After water passes through our homes, schools, and businesses, sewers collect this waste for disposal. Until the late nineteenth century, most waste emanating from cities was discharged directly into alleys and streets and eventually drained into nearby lakes and rivers without any treatment. In those days, you had to watch where you stepped. As the public health menace of these practices became apparent, communities began collecting and treating waste at specially constructed plants to render it less harmful to human health. These wastewater treatment plants are now critical features of urban development that enable the safe concentration of large numbers of people into relatively small urban areas. In addition to promoting public health and protecting the environment, these systems are part of the hidden infrastructure that supports economic growth and quality of life within our communities.[290]

WATERSHED OR DRAINAGE BASIN

John Wesley Powell famously proposed organizing political divisions in the western United States around the outlines of watershed boundaries.[291] Powell

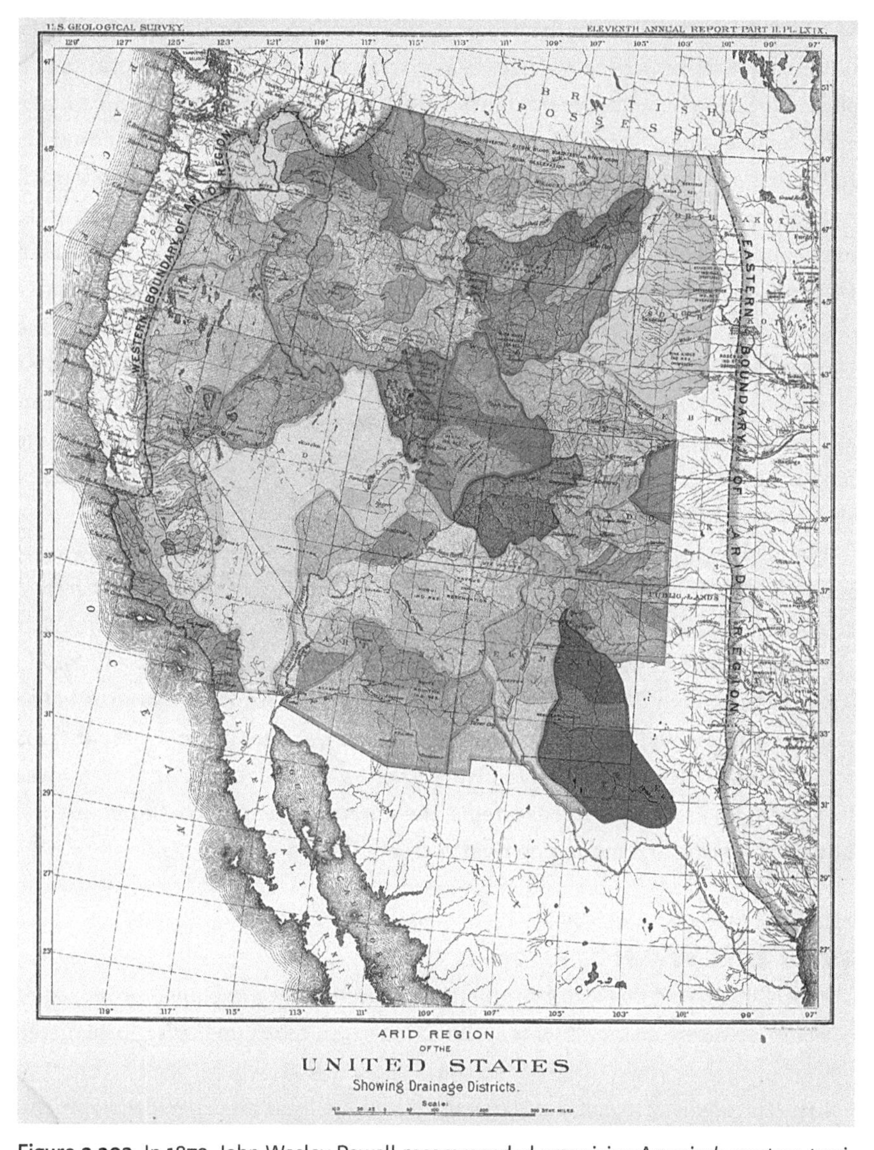

Figure 2.203. In 1879, John Wesley Powell recommended organizing America's western territories along watershed boundaries. It became one of his most famous proposals, but one that was never adopted. Here is a map of the watersheds in the western United States. Notice that the line stating "Eastern Boundary of Arid Region" coincides with the 100th meridian. Source: Eleventh Annual report of the Director of the United States Geological Survey, Part 2-Irrigation: 1889–1890, Annual Report 11.

believed that there was a natural logic in coordinating political boundaries with landscape features. These *watersheds* are land areas where all water drains to a shared location, such as rivers, lakes, or oceans. Although Powell's idea made considerable sense, the politicians who carved up the western lands ignored his recommendation. Eventually, some elements of Powell's proposal were adopted, not based on his writings, but because water users in the West demanded state oversight of water distribution. For this reason, dozens of drainage-basin based institutions, from water districts to watershed organizations and soil and water conservancy districts, use natural watershed outlines as their boundaries. One can easily determine watershed boundaries by tracing a line along the highest elevations between two areas on a map, often a ridge. Interestingly, large drainage basins, such as the Columbia River, contain thousands of smaller drainage basins nested within it. Each of these watersheds in turn may contain even smaller watersheds and so forth.

Water development has somewhat muddied the logic of natural watersheds. Today, multiple transbasin diversions move water from one watershed to another. Consider, for example, the Colorado River Aqueduct. This aqueduct carries approximately 1,200,000 acre-feet of water per year from the Colorado River to Southern California's Pacific coastal cities. Another example is the Roberts Tunnel that diverts Colorado River water under the Rocky Mountains to South Platte River and Denver. Nevertheless, although human waterworks have reengineered many river systems, the watershed remains the fundamental unit for studying and managing river systems.

WATER WAR

The term given to many water conflicts, *water war*, suggests struggles over access to water or distribution of water. The name often crops up during protracted disputes, such as in water grabs or court proceedings, where cities attempt to transfer water away from farms so that they might use it for urban development.

Westerners, particularly western journalists and writers, like to refer to conflicts over water as water wars but that overly dramatizes the reality. Water wars have reached the level of folklore in the region. These conflicts have entered popular literature (*The Milagro Beanfield War*) and movies (*Chinatown*). Water conflicts are highly emotional, creating tension and hard feelings, and are generally very expensive, lengthy, and litigious. Yet, they seldom rise to the level of violence, let alone general combat. The exception proving the rule is the Owens Valley water transfers. When Los Angeles completed the Los Angeles Aqueduct,

Figure 2.204. References to water wars occur in odd places throughout western culture, such as in this old postcard from Duckboy Cards Inc. of Hamilton, Montana. Image reproduced courtesy of Duckboy Cards Inc.

Owens Valley interests sabotaged the structure, but as far as we know, no violent injuries or deaths resulted.

Conflicts over water arise from various causes, including use, allocation, quality, and availability. Use conflicts may occur when, for example, irrigation diversions harm flows for recreation and habitat. When there are disagreements between states, river basins, or ditch systems, it is the allocation of the water that users contest. The availability of water in prior appropriation states may generate conflicts when water users disagree on the seniority of uses, the seniority system's fairness, or the kinds of beneficial uses authorized for the water. Pollution is another source of conflict: farmers complain that upstream industrial discharges harm their crops, or one nation complains that it is receiving salty water from another, just as Mexico has protested the saline Colorado River water it receives from California. Conflicts over water in the West will likely intensify in the future. However, because so much food is produced in water-rich areas and shipped to arid regions, it's unlikely that water shortages will cause starvation and revolt.[292] Human-induced climate change, population increases, habitat destruction, increases in economic inequality, and political polarization are additional factors that complicate water management decision making. These factors generate conflicts that will lead some in the media to say that water wars have erupted.

WATER WHEEL

Water wheels are an ancient technology that employs the energy of moving water to power various machines such as grist mills. People also use water wheels to lift water from a river into a ditch or an irrigated field. The Greeks, Romans, and ancient Chinese employed these wheels to move water. Because the efficiencies are rather low, water users seldomly use these wheels today. Three main varieties of water wheels exist, including the Norse wheel that turns millstones directly, overshot wheels that require an elevated stream to rotate the wheel, and undershot wheels that are employed to lift water. Vertical turbine pumps powered by electricity or other fuels are far more common today and can move greater quantities of water. Nevertheless, these old water wheels are quite scenic, even if they are inefficient by modern standards.[293]

Figure 2.205. This is an undershot water wheel that once irrigated fields along the Gunnison River of Colorado. Source: *Harpers Weekly* (1886).

WELL

People have been digging wells for millennia. In the Southwest, the earliest known wells are intricate hand-dug affairs. Archaeologists have discovered wells in sites such as the Anasazi ruin of Guaje Canyon in New Mexico (dating from the mid-1200s) and Paquimé in Northern Chihuahua, Mexico (its heydays

Figure 2.206. Large agricultural wells, such as this one near San Simon, AZ, supply irrigation water to nearby pecan groves. Notice the vertical turbine pump located on top of the wellbore situated just to the left of the large diesel engine that powers the pump. Photo courtesy of the author.

were in the 1300s and 1400s). The well at Paquimé was particularly impressive and contained rectangular stairways adorned with built-in shrines. The shrines demonstrate the importance of groundwater to the people who built the well. In the modern era, the West's aridity has given wells outsized significance in the region's water supply picture.

Accordingly, a considerable variety of wells exist that serve many purposes. Although any human-made shaft or hole dug into the earth used to obtain liquids is a *well*, the similarity between wells ends there. We mostly think of wells in terms of water supply or to produce gas or oil. For water purposes, people dig wells into consolidated or unconsolidated alluvium, bedrock (such as fractured granite), and other formations that produce water. Hydrologists also dig wells to collect groundwater data. These monitoring wells are employed for any number of reasons and range from recording aquifer levels to sampling groundwater under any site where pollution is suspected.[294]

WETLANDS

Because of its aridity, we rarely think of the West as an area rich in wetlands. Yet, these poorly drained sites in both fresh and saltwater settings are quite

Figure 2.207. Russell Lakes, shown here, is one of the highest alkaline wetlands in the world. Located within the Closed Basin area of Colorado's San Luis Valley, these created wetlands receive water from an artesian well. Russell Lakes provides an excellent migratory habitat for over one hundred bird species. The US Bureau of Reclamation developed these wetlands as partial compensation for other wetlands that it destroyed when constructing the Closed Basin Project. Photo courtesy of the author.

common. *Wetland* is a generic term used to describe many different kinds of wet habitats. Its use implies that the land is wet for some time, either permanently or intermittently. Other words used interchangeably with wetland include swamp, bog, cienega, marsh, and less commonly "marais" and "pocosin." Wetlands have numerous definitions and classifications stemming from both regulatory and scientific needs. Most definitions require the presence of three features for a site to be considered a wetland. These features include a hydrologic aspect where there is a high degree of flooding or soil saturation, vegetation that is adapted to grow either in water or soils that remain saturated for long periods, and soils that show signs of oxygen deficiency due to saturation. Look for wetlands where the local topography allows water to collect on the ground or where groundwater discharges to the surface. The result of both is that a site becomes saturated with water for extended periods.

People use a wide variety of names to describe the topographic setting where you may find wetlands, such as depressions, swales, sloughs, prairie potholes, playas, vernal pools, oxbows, and glacial kettles. Likewise, the relatively low-lying areas subject to flooding that contain wetlands include intertidal flats and marshes, coastal lowlands, sheltered embayments, deltas, and flood plains. Broad flat areas without clear outlets or drainage also support wetlands and include stream divides and permafrost muskegs. On sloping terrain below springs and seeps or relatively flat or sloping areas adjacent to bogs, the land can remain saturated and support wetlands.[295]

Wetlands perform valuable functions in the environment and for people. In the most simplistic terms, wetlands provide water quality as well as hydrologic and habitat functions. Some of these functions include mitigation of floods, water storage, nutrient assimilation and transformation, and supporting a wide diversity of plants and animals. Although wetlands vary widely in the quality of services they provide, wetlands generally contain some of the most productive habitats in the world. They also provide direct benefits to people through hunting, fishing, and bird watching. Humans have affected wetlands in fundamental ways through outright filling or dewatering, grazing, clearing, dredging, and the introduction of exotic species. Elsewhere, people have created wetlands by moving water around the western landscape through water management practices (ditch digging and irrigation, for example) or construction of reservoirs. Natural variations of climate and drought also affect wetlands and their function during any given year.

WET WATER / PAPER WATER

The actual flow in a river or ditch is known as the "wet water" in a system. This flow includes all the water: streamflow, ditch flows, reservoir and groundwater levels, and so forth. "Paper water," on the other hand, is the theoretical quantities of water found in water right decrees or administrative rulings. For example, a ditch decree might be for 25 cfs, but if the flow in the stream from which that ditch diverts is only 15 cfs, then there is a 10 cfs difference between the wet water and paper water in the system. The imbalance between the two is often cause for great contention between users. Indeed, California courts distinguish between wet water and paper water: "three cases clarified that 'wet water' is water supply that is actually and physically available for diversion and use, while 'paper water' are [*sic*] water supply entitlements that are referenced in documents . . . but that do not in fact exist due to hydrological realities, environmental restraints, or uncompleted infrastructure." The court elaborated, explaining, "Entitlements represent nothing more than hopes, expectations, water futures or, as the parties refer to them, 'paper water.' Paper water always was an illusion. 'Entitlements' is

Figure 2.208. Wet water is the actual water flowing in a stream or ditch. Paper water only exists as an artifact of a water rights decree or contract. Photo courtesy of the author.

a misnomer, for contractors surely cannot be entitled to water nature refuses to provide or the body politic refuses to harvest, store and deliver."[296]

Furthermore, recent water rights decrees often include values for small quantities of water (somewhat arcane things including evaporation replacements, stream loss factors, ditch seepage factors, and others that have importance only to hydrologists and attorneys). These values can be very tiny and create measurement issues for those tasked with administering water rights. Consider the situation where a water rights decree directs a water manager to release a tiny quantity of water, let's say 0.01 cfs for discussion purposes. If a stream or ditch is flowing at several hundred cfs and the nearby gage can only measure in the tens of cfs, it becomes practically impossible to measure the actual water quantities that the water rights decree demands be released. In such cases, these minuscule amounts of water are, for all intents and purposes, paper water.

WHOOPING CRANE

The *whooping crane* (*Grus americana*) is North America's tallest bird. Sadly, it is also one of the most imperiled. Males stand nearly 5 feet tall when fully erect. Adult whooping cranes are known for their snowy white feathers and black

Figure 2.209. The whooping crane is one of the iconic megafaunas of the American West. Fights over water to ensure its survival have led to epic clashes between environmentalists and water developers. Shown here is a whooping crane (flying) with sandhill cranes. Photograph by Jim Hudgins courtesy of the US Fish and Wildlife Service. Public domain image.

primary feathers on the wings. They also sport a bare red face and crown. Their name likely originates from the loud, single-note cry that they repeat when alarmed. Whoopers, as they are commonly called, can live as long as thirty years. But they seldom achieve this age in the wild given the large number of threats that the bird faces. According to the US Fish and Wildlife Service, whooping cranes currently live in three locations in the wild and at twelve captive sites. As of July 2010, the total wild population stood at only 383 individuals.

A central flyway, known as the Aransas–Wood Buffalo flyway, extends from the Aransas National Wildlife Refuge in Texas to the Wood Buffalo National Park in northeastern Alberta, Canada. There used to be a Rocky Mountain flyway for the bird, but the last whooper died in the spring of 2002. Around 150 birds live in various zoos. Because of their imperiled status, the US Fish and Wildlife Service listed whoopers as threatened under the Endangered Species Act. This threatened status requires the US government to consider how its decisions affect whooping cranes and their habitat. Since the choke point in the whooping crane's migration is along the Platte River in central Nebraska, water development actions upstream from this location in Nebraska, Colorado, and Wyoming are subject to evaluation and modification. Should the US Fish and Wildlife Service determine that even minuscule developments or diversions affect the bird's habitat, the water developers must modify their project and pay toward whooping crane protection efforts. Given its protected status, environmentalists have made the whooping crane a symbol for conservation, while water developers have derided conservation efforts that they believe interfere with water development.[297]

WINDMILL

People have employed windmills to grind grain and pump water since their introduction in Persia sometime between AD 500 and AD 900. The Chinese refined windmill technology, introducing vertical axis turbines in the early 1200s. By the late 1200s, Western Europeans were building them, and in the late 1300s, the Dutch began constructing multistory towers that became the signature architectural feature of that nation. Over time, improvements in mechanical efficiency and sail design dramatically streamlined the operation of these machines. In the nineteenth century, pumps driven by steam generation and later electrical power reduced the direct use of windmills. However, in 1854, Daniel Halladay of Connecticut patented a self-governing windmill—one that could change direction with changing winds—that was capable of drawing water from as deep as 28 feet. In the arid West, this was a game-changer for farmers and ranchers who could now supply water to sites otherwise devoid of water. Halladay's invention quickly spread, providing water for crops and livestock across the Great Plains and Mountain West. In time, the efficiency of the Halladay windmill increased, and variants of this invention remain in everyday use into the twenty-first century.

Figure 2.210. A Halladay style windmill feeds groundwater to a stock pond on Navajo Nation land in western New Mexico. Photo courtesy of the author.

YELLOWSTONE LAKE

A series of major volcanic episodes created the remarkable geologic setting that we now preserve as Yellowstone National Park. This park is famous for its geysers, hot springs, fumaroles, and other geothermal phenomena and is nearly unsurpassed worldwide for this activity. Photographs by William Henry Jackson, paintings by Thomas Moran, and geologist Ferdinand Vandeveer Hayden's reports inspired Congress to create Yellowstone National Park in 1872. Yellowstone's water features, including its thermal springs and geysers, Yellowstone River's falls, and Yellowstone Lake form the heart of America's first national park. The Yellowstone area emerged as a unique geological region through a series of gigantic volcanic eruptions about 2.2 million to 600,000 years ago. About 600,000 years ago, a catastrophic outpouring of rhyolite, pumice, and ash covered more than 900 square km of the greater Yellowstone area during the most recent eruption. As the eruption subsided, a ring of fracture zones collapsed to form a caldera that created the basin in which Yellowstone Lake formed. During the Pleistocene, glacial activity further modified the Yellowstone area. With the retreat of the last glaciers about twelve thousand years ago, the Yellowstone region, including Yellowstone Lake, came to look much as it does today.

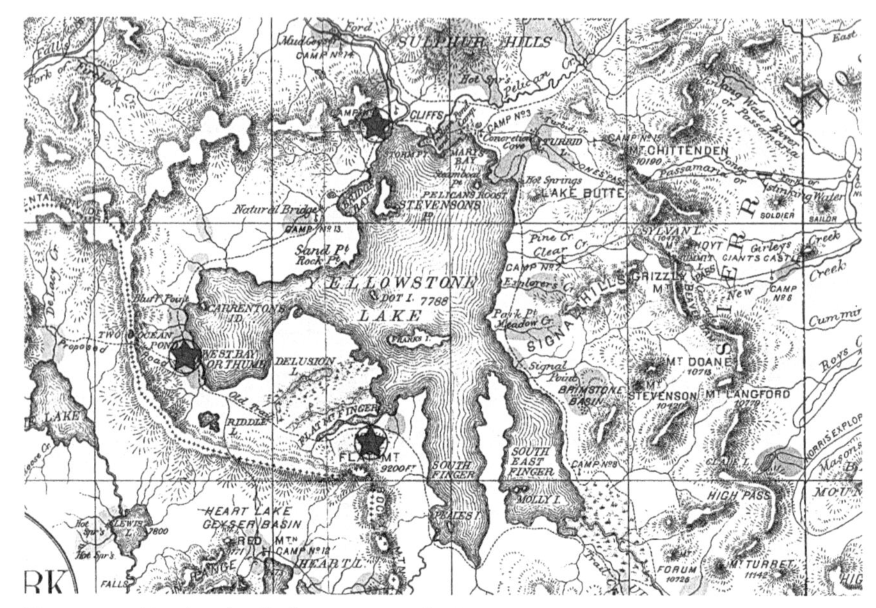

Figure 2.211. Here is a detail of an 1881 map of Yellowstone National Park showing Yellowstone Lake. Although the geology of the region was poorly understood at the time, the general outline of the Yellowstone Caldera is clearly visible as a broken ring of hills that bounds the lake. Public domain image courtesy of the Library of Congress.

In the Yellowstone region, fewer than one half of the lakes and streams contained fish at the time Europeans first arrived. However, between 1881 and 1955, fishermen released some 310 million fish into the water bodies within Yellowstone National Park. Yellowstone Lake was ground zero for these efforts, and this has led to dramatic ecological changes in its waters. Its thirteen native fish species have found themselves defending against introduced predator species and have wound up hybridizing with others. These introductions have led to the decline of the endemic ecology in Yellowstone as new species become naturalized alongside the old.[298]

YIELD

Yield is an engineering term describing a volume of something produced over a given period of time. This seemingly arcane term is central to many contemporary conversations about western water management. In particular, debates over the meaning of yield, and the declared yield of water projects, have crucial consequences for sizing dams, claiming water rights, estimating water project

Figure 2.212. Here, the author inspects the output of a high-capacity irrigation well in Arizona's Gila River Valley. Photograph courtesy of Tina Tan.

resilience, and calculating the impacts of droughts and climate change. Yield is necessary for determining the cost-benefit of water projects. Debates over this figure's accuracy often lie at the center of fights over dam authorization and water rights litigation.

In water management, engineers usually report yield as a quantity of water produced over a specific period, such as gallons per minute (as in wells or domestic settings) or cubic feet per second (in farms and industrial environments). At first glance, yield appears to be a constant or immutable value, but in reality it may vary quite dramatically over time. The yield of wells, in particular, can diminish over time if the natural recharge of the aquifer is less than the amount of water pumped from that aquifer. Likewise, the project yield of municipal systems may vary (sometimes significantly) from year to year depending on variations in local rain or snowfall, aging infrastructure (in particular siltation of reservoirs), climate change and global warming, or from challenges to water rights.

Often water managers use a related and particularly murky term, "firm yield," to describe expected project deliveries from municipal water supply projects. Specifically, firm yield means the amount of water a project can produce during a "design drought." Water resource planners measure the drought as the time that the system (usually a reservoir) is last full and spilling to the time that it is completely empty and begins to refill. For example, if a design drought is ten years, and a dam can store 10,000 acre-feet of water, the firm yield of that dam is given as 1,000 acre-feet per year. Figuring out what a design drought looks like can take some wizardry by way of lots of assumptions on climate, weather, how much the past reflects the future, how the watershed has evolved over the years (such as fires and deforestation), water rights considerations, how other water projects affect the system, and so on. In essence, the projected yield of a system depends on how reasonable the assumptions that went into it are. In recent years, the accelerating impacts of climate change have called many old beliefs into question. Water users are learning the hard way that system reliability is not what it used to be. Here more than almost anywhere else in water management, the adage about junk into a calculation means junk out is remarkably prescient. The problem is that firm yield numbers sound, well, firm, and so mask all the uncertainty that goes into these calculations.[299]

ZYBACH, FRANK, AND CENTER PIVOT IRRIGATION

In 1947, Frank Zybach, a farmer from Columbus, Nebraska, saw an irrigation demonstration on a farm in Strasburg, Colorado. Workers scurried back and forth to move irrigation pipes that were fitted with sprinkler heads on posts. Although these pipes were more efficient than flood irrigation and were especially useful on hilly terrain, Zybach thought that there had to be a better way than having workers assemble and disassemble everything whenever they wanted to flood a different area. Zybach, a long-time tinkerer, began thinking about how to improve irrigation machinery. Setting his mind to the problem, Zybach took about a year to develop the "center pivot irrigation system." His design consisted of a line of pipes pivoted around a wellhead or pump that supplied the water. In Zybach's first iteration, evenly spaced towers moved about on skids and utilized guy wires that extended from the towers to sections of pipe that had sprinklers attached. He soon replaced the skids with wheels and powered the system with water pressure from the pump. By 1949, Zybach had submitted a patent application for the "Zybach Self-Propelled Sprinkling Apparatus."[300]

Center pivot irrigation systems became an instant success and revolutionized the irrigation industry and, at the same time, reordered the geography of the high plains as farmers recontoured their fields to take advantage of this new technology. Facilitating the growth of these systems was the New Deal program of government-subsidized rural electrification that supplied power to all the

Figure 2.213. Here, a typical center pivot irrigation system waters land located in Colorado's San Luis Valley. Photo courtesy of the author.

pumps at the center of the pivot. Within thirty years of Zybach's patent, farmers were irrigating more than 12 million acres of land with center pivots. Modern systems commonly employ electric motors that propel the pipes and use GPS sensors coupled to variable-rate sprinkler heads to ensure even water distribution. With the adoption of center pivots, groundwater exploitation took off. By efficiently pumping groundwater aquifers across the West, the water table in many aquifers has declined, sometimes dramatically. Perhaps most notorious is the Ogallala of the High Plains, where the water table has gone down as much as several feet per year. Ironically, Zybach's invention, once heralded as a transformative technology, has directly brought an end to irrigation in some of the very areas it meant to serve.[301]

PART 3

THINKING ABOUT WESTERN WATER IN THE TWENTY-FIRST CENTURY

The more one contemplates water in the West, the more it becomes apparent that so many seemingly stand-alone topics actually have deep interpenetrating connections. Having read this far, you surely noticed by now how multiple entries seem to allude to other topics: Geology and climate affect hydrology and biology, and that in turn impacts how we adapt to the environment, the course of development, and the economic decisions made by cities. Climate change affects everything. And we affect the climate. People depend on freshwater ecosystems that we intentionally or unintentionally alter. We've even created hybrid freshwater ecosystems on formerly dry lands. We preserve some of these ecosystems as National Wildlife Refuges or local natural areas. Introduced species impact hydrology, geomorphology, and endangered species. Ironically, some endangered species rely on introduced species for their habitat. Historical events continue to have significant ramifications for contemporary communities. Art and culture influence societal perceptions regarding water and these perceptions influence political decisions about water development and the environment. It's tempting to go on and on about connections and feedback loops, but I think you get the idea.

With all these connections in mind, I wish to offer a few parting thoughts and observations about western water. As I write these words, my eyes water from the thick acrid smoke emanating from wildfires burning several miles from my home in Boulder, Colorado. We've had some major fire seasons in previous years, but this one stood out above all others in its destructiveness. In 2002, the Hayman fire consumed more than 138,000 acres southwest of Denver, becoming what was at that time the largest wildfire in Colorado state history. The record stood until this summer when lightning ignited the Pine Gulch fire near Grand Junction, Colorado, burning more than 139,000 acres to exceed the Hayman record. On the heels of that was the Cameron Peak fire that started west of Fort Collins, Colorado. Burning from late summer and into the fall, the Cameron Peak fire consumed more than 207,000 acres giving it the dubious distinction of becoming the largest wildfire in Colorado history. Then, the East Troublesome fire violently erupted more than fivefold in area to 192,000 acres to become the second-largest fire in Colorado's history. Therefore, in the year 2020, Colorado has experienced the state's three largest fires in recorded history in just three months.

Seeing the volcanic cloud-like plumes of smoke pouring across the Continental Divide from the East Troublesome fire, I can't help but dread the future climate change disasters awaiting us. If the Hayman fire is any indication, these burns will substantially damage the environment and adversely impact the people who rely on the affected watersheds. After Hayman, for example, Denver Water spent $27 million on water quality treatment, sediment and debris removal, reclamation, and infrastructure projects within its water collection system to recover from the fire.[302] And that does not include all the other federal, state, and local resources that went into recovery or the dollars spent to replace the private property lost.

Our rivers and their ecosystems, water users, and the general public stand to suffer long-term damage from the fires of 2020 alone. And this is just in Colorado. The situation in California and other western states is equally distressing. Wildfires burned well over 4 million acres in California during 2020, and 5 of the top 20 most massive wildfires in the state's history happened this year. The year 2020 was also one of the most destructive fire seasons on record for Washington, Oregon, and Arizona. Furthermore, as a backdrop to all these fires, the US Drought Monitor reports that much of the western United States is undergoing moderate to exceptional drought.

Climate change threatens many of the things I describe in this book. Current research suggests that between 2000 and 2022 the southwestern United States is undergoing the driest 22-year period since 800 AD.[303] Reduced flows in rivers, watershed damage, debris flows in the aftermath of wildfires, water quality impacts, diminishment of recreation and fishing opportunities, water supply uncertainty, and receding glaciers are just some of the effects. The list goes on. All the things we value and need within our water systems seem at risk.

Discussions about biodiversity in the twenty-first century need to evolve too. We can no longer set aside discrete blocks of land and expect that the species residing there are secure. With a changing climate comes geographical shifts in a whole host of species. Designating new protected lands that can accept climate refugees might become one of the pressing needs for land and water conservation in the twenty-first century. Take the desert pupfish. As the climate warms and groundwater pumping in the basins that feed this fish's habitat continues, biologists will need to get inventive to prevent extinction. And while the Endangered Species Act is America's foremost law to protect threatened and endangered species, it is an imperfect tool for safeguarding ecosystems. Rare birds like the southwestern willow flycatcher and the whooping crane can force water management changes within watersheds. Still, the Endangered Species Act does a weak job of looking at watersheds from a comprehensive perspective to keep unprotected species from decline. Likewise, it's essential to maintain resiliency and flexibility in managing rare and endangered species if we are to protect the many species that live in the West's riparian corridors.

In managing our water resources, westerners created vast hybrid ecosystems that simultaneously intermingle endemic and introduced species, affect biological processes and human activities, alter hydrology, and modify the physical environment. Many of these ecosystems are protected and provide sanctuary for a long list of species. A good deal more of these lands remain unprotected, even though they too are home to many valuable species. How we manage these hybrid ecosystems and the river flows and groundwater supplies that support them will largely determine which endemic species survive, which invasive species are controlled or naturalized, and what wildlands, wetlands, and rivers will persist for future generations. Increasingly, land and water managers will need to incorporate these hybrid ecosystems into biodiversity conservation strategies.

Throughout the history of western water, we have seen that relying on pseudoscience has led to some tragic outcomes. As a case in point, the debunked notion that "rain follows the plow" worked out well for boosters and real estate speculators, but in the end, many farm families paid the toll through drought, bankruptcies, and failed dreams. I see echoes of "rain follows the plow" in our current climate debates. Like nineteenth-century real estate speculators, it's common to see those who profit from carbon emissions questions or dismiss scientific evidence to the contrary. In the meantime, westerners stare into a potentially bleak and costly climate future. We also find the hand of pseudoscience and misinformation at work with those who will debate the efficacy of using fluorides to prevent tooth decay. Conversely, many people happily utilize water dowsing instead of a hydrologic analysis that would prove far more reliable. There is great potential for disaster or loss if we rely on pseudoscience and not carefully vetted scientific theories backed by empirical data.

In the West, climate uncertainty is nothing new. For thousands of years, people have responded to aridity and climate change. What's different today is that the rate of change has dramatically accelerated, and we must act now to make our water systems resilient in the face of rapidly shifting conditions. Fortunately, many large municipalities are developing plans to address the changing climate. However, tens of millions of people now live in this once sparsely populated region. Although many millions live in the West, national and international trade networks in food and products might insulate us from some of the worst human impacts of aridity. Because so much food and goods come from water-rich regions and get shipped into the West as virtual water, climate change impacts do not affect us as severely as it would have in the past. After all, residents of counties experiencing exceptional drought would be far worse off if they did not get their food elsewhere. However, our large population does constrain the choices for managing water and reduces the resiliency within our water systems to respond to aridification.

Nevertheless, by nature, I am an optimist. Although I find many of the problems facing us quite dispiriting, I believe that if we muster the political will to alter course, we can head off some of the worst effects of climate change. But our window to act narrows by the day. Fortunately, the work of Elinor Ostrom shows us that we do not have to face a "tragedy of the commons" as we go about managing our water resources. Rather, history proves that practicing cooperation, collaboration, and communication provides a path to head off or ameliorate the thorniest problems. Political grandstanding, scoring points, taking entrenched positions will not give us a better West. Negotiations and compromise to find workable solutions are never easy or sexy, but they can and do succeed.

America owes a vast debt of gratitude to its Indigenous and Hispanic people who were the true pioneers in understanding and managing water resources in the West. Euro-Americans were Johnny-come-latelies who came, saw, replicated, and then built upon the traditional knowledge received from the Indigenous and Hispanic people. Unfortunately, the Euro-Americans also colonized the West by marginalizing or eliminating many of these same people. Embedded in the Prior Appropriation Doctrine is the legacy of American colonialism. The idea of "first in time" is a colonial construct that ignored the human rights of all the people who preceded the Euro-American occupation. It wasn't until Euro-Americans took most of the water resources for themselves that they grudgingly provided water to Indigenous people (after the US Supreme Court rendered the Winters decision) within the Prior Appropriation Doctrine framework.

Moreover, as colonizers, Euro-Americans recognized the land and water rights of Hispanic peoples to lessen the possibility for an insurrection in those communities. To this day, many Indigenous and rural Hispanic communities have underfunded, substandard water infrastructure—legacies of the colonial past. And, as the Standing Rock Sioux discovered to their despair, Indigenous communities still find themselves marginalized as federal agencies make decisions affecting their welfare. Ending this paternalism and reconciling these colonial legacies will benefit all western residents.

And although Euro-American colonists in the West initially used the idea of riparian rights that came from European common law, they mostly rejected it when they realized that this system could serve to monopolize resources to the detriment of the broader good. In the West, only California, Oregon, Washington—states that are less arid than the other western states—retained elements of the Riparian Doctrine. Consequently, these three states ended up with a hybrid water rights system that we commonly call the California Doctrine. It's healthy to keep in mind that a central premise for justifying the Prior Appropriation Doctrine was to benefit the maximum number of people. For me, that means not just irrigating or serving people potable water but also providing everyone access to clean rivers and lakes for recreation, fishing, and contemplation.

Another takeaway is how much literature, art, and cinema—popular culture—heavily influence how Americans understand and perceive western water. We've seen how one book by a gifted author like Edward Abbey or Norman Maclean might inspire a whole generation of young people to oppose dam construction or pick up fly rods. Likewise, cinema can shape perceptions about corruption and water development. Art, exemplified by Oskar Hansen's Egyptian-inspired statues, supports notions of empire. Elsewhere, conceptual art, such as Robert Smithson's *Spiral Jetty* encourages us to consider the environment in new and interesting ways. Similarly, stunts, such as moving the London Bridge to the Arizona desert, or building an elaborate water feature like that in front of the Bellagio in Las Vegas, help create the illusion that people can tap vast water supplies to provide all the water needed to thrive in a desert oasis. Photography, too, has the powerful capability to mold opinions. Photographs simultaneously document an era and influence future generations' judgments of those past times. And when we speak to each other, it's easy to forget how many words that we use to describe the West and its landscape originally came from Spain and Mexico. Moreover, large dam and canal projects of the twentieth century took place within the framework of American modernity, capitalism, and empire in the West. As such, these water projects comprise a central part of the West's material and intellectual culture.

When I went to school for geology, one of the core mantras in the field was that by studying the present, one could unlock the past's mysteries. But after spending a career in water resources, I like to invert that wisdom and posit that to understand western water better, we must study our past to gain deeper understandings of the present. What I mean by this is that we need to continue to investigate how the Indigenous, Hispanic peoples, and the early Euro-Americans changed the land and rivers and adapted to the western environment, and consider what motivated their decision making. Understanding the adaptations that they made to live in this arid environment can inform our contemporary decision making. Furthermore, it's important to continue studying the geological, hydrological, and biological systems that we depend on. Our cultural and natural systems are intricate and tightly entangled. Water management is complex, but not hopelessly complex. We can get it done.

NOTES

1. For a detailed project history, see Garrit Voggesser, *The Dolores Project*, Bureau of Reclamation History, US Bureau of Reclamation, https://www.usbr.gov/projects/pdf.php?id=113. Also, for more general background information on the Colorado River Storage Act, Wayne Aspinall's role, and how the Dolores Project fits into all this, see Marc Reisner, *Cadillac Desert: The American West and Its Disappearing Water* (New York: Penguin, 1993), 293–98.

2. After declaring this feeble project benefit, the US Bureau of Reclamation apologetically adds, "Sorry—No additional information is available at this time." See Voggesser, *Dolores Project*.

3. The construction and operational costs come from Voggesser, *The Dolores Project*. It is expected that power revenues, assessments from local project water users, and landowner taxes will repay the costs of the Dolores River Project.

4. Roland Barthes, *Camera Lucida: Reflections on Photography* (New York: Macmillan, 1981), 80.

5. Many photography books investigate our interaction with river systems. Two notable works that deal specifically with western water were particularly influential. Eliot Porter, *The Place No One Knew: Glen Canyon on the Colorado* (Oakland, CA: Sierra Club, 1966). This book helped unveil to the world the amazing scenic resources that were lost with the flooding of Glen Canyon. Also H. G. Stephens, *In the Footsteps of John Wesley Powell: An Album of Comparative Photographs of the Green and Colorado Rivers, 1871–72 and 1968* (Chicago: Johnson Books, 1987). Stephens's book of comparative images helped make a broad audience of people aware of how much western landscapes have changed due to water development.

6. Jacobs's quote from footnote 5 in William Deverell, "Fighting Words: The Significance of the American West in the History of the United States," *Western Historical Quarterly* 25, no. 2 (1994): 185–206.

7. For example, see Lauriane Bourgeon, Ariane Burke, and Thomas Higham, "Earliest Human Presence in North America Dated to the Last Glacial Maximum: New Radiocarbon Dates from Bluefish Caves, Canada," *Plos one* 12, no. 1 (2017): e0169486.

8. Ned Norris Jr., "Stop Destroying Tohono O'odham Sacred Sites," *High Country News* (December 2020): 43.

9. For more about the impact of Spain, Mexico, and the United States on the Indigenous people of the Southwest and how this region eventually emerged as a region of the United States, see Edward H. Spicer, *Cycles of Conquest: The Impact of Spain, Mexico, and the United States on the Indians of the Southwest, 1533–1960* (Tucson: University of Arizona Press, 1962).

10. William Cronon, *Nature's Metropolis: Chicago and the Great West* (New York: W. W. Norton and Company, 2009).

11. Specifically, Powell used the *isohyetal*, or mean annual rainfall line, of 20 inches to distinguish wetter areas to the east from drier regions to the west. See John Wesley Powell, *Report on the Lands of the Arid Region of the United States: With a More Detailed Account of the Lands of Utah. With Maps* (Washington, DC: US Government Printing Office, 1879), 2.

12. Historian Richard White talks about the problems of defining the West in *A New History of the American West* (Norman: University of Oklahoma Press, 1991).

13. Even though I will use the 100th meridian to demarcate the humid East from the arid West, when I talk about western water law or river access, I will confine those discussions to the contiguous United States west of the eastern boundaries of Montana, Wyoming, Colorado, and New Mexico.

14. To learn more about this colonial history, I recommend Patricia Nelson Limerick, *The Legacy of Conquest: The Unbroken Past of the American West* (New York: W. W. Norton and Company, 1987); also White, *A New History of the American West* (Norman: University of Oklahoma Press, 1991).

15. See Robert R. Crifasi, *A Land Made from Water: Appropriation and the Evolution of Colorado's Landscape, Ditches, and Water Institutions* (Boulder: University Press of Colorado, 2015), chap. 6.

16. James A. Crutchfield, *Revolt at Taos: The New Mexican and Indian Insurrection of 1847* (Yardley, PA: Westholme Publishing, 2015).

17. Article VIII, Treaty of Guadalupe Hidalgo (1848), https://www.ourdocuments.gov/doc.php?flash=false&doc=26&page=transcript.

18. Patricia Nelson Limerick, Jeffery Hickey, and Richard DiNucci, *What's in a Name? Nichols Hall* (Boulder: University Press of Colorado, 1987).

19. David H. Getches, *Water Law in a Nutshell* (No. 346.0432 G394 2009) (St. Paul, MN: Thomson West, 2009), 14.

20. For a good primer on water rights, see Getches, *Water Law in a Nutshell* (No. 346.0432 G394 2009) (Saint Paul, MN: Thomson West, 2009).

21. The other states with a hybrid water rights system lie east of the 100th meridian, including Kansas, Mississippi, Nebraska, North Dakota, Oklahoma, South Dakota, and Texas.

22. There are a number of excellent books that investigate and analyze the course of water development in the American West. Two that I highly recommend are Marc Reisner,

Cadillac Desert: The American West and Its Disappearing Water (New York: Penguin, 1993); and Donald Worster, *Rivers of Empire: Water, Aridity, and the Growth of the American West* (Oxford: Oxford University Press, 1992).

23. Geographers are particularly good at mapping the interactions between our cultural and physical environments. A branch of geography known as "political ecology" investigates relationships, interactions, and outcomes between politics, economics, society, and the environment. As such, this discipline is especially useful for studying western water. Thinking about western water from the perspective of political ecology inspired the organizational outline of this book. For more about political ecology, see Thomas E. Sheridan, "Arizona: Political Ecology of a Desert State," *Journal of Political Ecology* 2 (1995): 41–57; Erik Swyngedouw, "Modernity and Hybridity: Nature, Regeneracionismo, and the Production of the Spanish Waterscape, 1890–1930," *Annals of the Association of American Geographers* 89, no. 3 (1999): 443–465. http://dx.doi.org/10.1111/0004-5608.00157; and Robert R. Crifasi, "Political Ecology of Water Use and Development," *Water International* 27, no. 4 (2002): 492–503. http://dx.doi.org/10.1080/02508060208687037.

24. Edward Abbey, *The Monkey Wrench Gang* (Salt Lake City: Dream Garden Press, 1985), 112.

25. "Shadows from the Big Woods," in Edward Abbey, *The Journey Home: Some Words in the Defense of the American West* (New York: E. P. Dutton, 1977), 223.

26. Edward Abbey, *A Voice Crying in the Wilderness (Vox Clamantis in Deserto): Notes from a Secret Journal* (New York: St. Martin's Griffin, 1989), 97.

27. Edward Abbey, *Down the River* (New York: Plume, 1991), 201.

28. If you want to get a sense of who Ed Abbey was, read his books first. For a biography of Abbey, see J. M. Cahalan, *Edward Abbey: A Life* (Tucson: University of Arizona Press, 2003).

29. To delve into what life is like for people who use acequias, see Stanley G. Crawford. *Mayordomo: Chronicle of an Acequia in Northern New Mexico* (Albuquerque: University of New Mexico Press, 1993). See also Sylvia Rodríguez, *Acequia: Water Sharing, Sanctity, and Place* (Santa Fe, New Mexico: School for Advanced Research Resident Scholar Book, 2006). For an elegant description of life in rural New Mexico and along acequias, see William deBuys and Alex Harris, *River of Traps: A Village Life* (Albuquerque: University of New Mexico Press, 1990). For an overview of acequia institutions, see Charlotte Benson Crossland, "Acequia Rights in Law and Tradition," *Journal of the Southwest* (1990): 278–87.

30. Donald Worster, *A River Running West: The Life of John Wesley Powell* (New York: Oxford University Press, 2002), 477.

31. Ed Quillen, "What Size Shoe Does an Acre-Foot Wear?" *High Country News*, November 10, 1986, 16–25.

32. David Owen, "The World Is Running Out of Sand," *New Yorker*, May 29, 2017.

33. Robert R. Crifasi, "Reflections in a Stock Pond: Are Anthropogenically Derived Freshwater Ecosystems Natural, Artificial, Or Something Else?" *Environmental Management* 36, no. 5 (2005): 625–639.

34. Louis Menand, "Out of Bethlehem: The Radicalization of Joan Didion," *New Yorker*, August 17, 2015.

35. For more on agriculture, see Julie Guthman, *Agrarian Dreams: The Paradox of Organic Farming in California*, vol. 11 (Oakland: University of California Press, 2014); April R. Summit, *Contested Waters: An Environmental History of the Colorado River* (Boulder: University Press of Colorado, 2012), chap 2.

36. J. S. Gates, *Ground Water in the Great Basin Part of the Basin and Range Province, Western Utah*, in Cenozoic Geology of Western Utah: Sites for Precious Metal and Hydrocarbon Accumulations (Utah Geological Association, 1987), 75–90.

37. D. T. Patten, L. Rouse, and J. C. Stromberg, "Isolated Spring Wetlands," in The Great Basin and Mojave Deserts, USA: Potential Response of Vegetation to Groundwater Withdrawal. *Environmental Management*, 41, no. 3 (2008): 398–413.

38. Thomas Dunne and Luna B. Leopold, *Water in Environmental Planning* (New York: W. H. Freeman and Company, 1978), 194.

39. MOMA Learning, "What Is Modern Art?" https://www.moma.org/learn/moma_learning/themes/what-is-modern-art/.

40. Janet Haven, University of Virginia American Studies Program and the 1930's Project. http://xroads.virginia.edu/~ma98/haven/hoover/modern.html.

41. W. R. Osterkamp, *Annotated Definitions of Selected Geomorphic Terms*, USGS Open-File Report 2008–1217 (2008).

42. T. Mitchell Prudden, "The Prehistoric Ruins of the San Juan Watershed in Utah, Arizona, Colorado, and New Mexico," *American Anthropologist* 5, no. 2 (1903): 224–288.

43. Samuel Franklin Emmons, Whitman Cross, and George Homans Eldridge, *Geology of the Denver Basin in Colorado*, USGS Monograph Vol. 27. US Government Printing Office, 1896.

44. For more on artesian wells see F. G. Driscoll, *Groundwater and Wells, 2nd ed.* (St. Paul, MN: Johnson Division, 1986).

45. For a couple of biographies on Wayne Aspinall see S. C. Schulte, *Wayne Aspinall and the Shaping of the American West* (Boulder: University Press of Colorado, 2002). Also S. C. Sturgeon, *The Politics of Western Water: The Congressional Career of Wayne Aspinall* (Tucson: University of Arizona Press, 2002).

46. For more on augmentation see Colorado Division of Water Resources, https://dwr.colorado.gov/services/water-administration/augmentation-plans. Last accessed October 28, 2020.

47. W. R. Osterkamp. *Annotated Definitions of Selected Geomorphic Terms*, USGS Open-File Report 2008–1217. (2008). https://www.wou.edu/las/physci/taylor/g322/Osterkamp_2008_USGS_ofr20081217.pdf.

48. For more on avalanches see Avalanche.org, https://avalanche.org; and Colorado Avalanche Information Center, https://www.avalanche.state.co.us.

49. M. Jane Young, *Signs from the Ancestors: Zuni Cultural Symbolism and Perceptions of Rock Art* (Albuquerque: University of New Mexico Press, 1988), 174.

50. Jonathan Bailey, *Rock Art: A Vision of a Vanishing Cultural Landscape* (Chicago: Johnson Books, 2016), 71.

51. Discharge is the volume of water moving in the channel at specific times.

52. W. R. Osterkamp, *Annotated Definitions of Selected Geomorphic Terms*, USGS Open-File Report 2008–1217 (2008).

53. W. R. Osterkamp, *Annotated Definitions of Selected Geomorphic Terms*, USGS Open-File Report 2008–1217 (2008).

54. Geomorphologists classify bars based on shape and location within a river channel: transverse, longitudinal, point, etc.

55. For more on baseflow see Thomas Dunne and Luna B. Leopold, *Water in Environmental Planning* (New York: W. H. Freeman and Company, 1978), 288.

56. Bryan Walsh, "Dying for a Drink," *TIME*, December 4, 2008. http://content.time.com/time/magazine/article/0,9171,1864440,00.html. Accessed October 28, 2020.

57. If you want to read more about beavers and their impacts on geomorphology, see chap. 2 in Ellen Wohl's informative book, *Virtual Rivers, Lessons from the Mountain Rivers of the Colorado Front Range* (New Haven, CT: Yale University Press, 2001).

58. For more on the fur trade, John Jacob Astor, and the rise of the American empire, see Eric J. Dolin, *Fur, Fortune, and Empire: The Epic History of the Fur Trade in America* (London: W. W. Norton, 2010).

59. For more on beneficial use see D. H. Getches, *Water Law in a Nutshell* (No. 346.0432 G394 2009), Thomson West, 2009.

60. Cristina M. Villanueva, Marianna Garfí, Carles Milà, Sergio Olmos, Ivet Ferrer, Cathryn Tonne, "Health and Environmental Impacts of Drinking Water Choices in Barcelona, Spain: A Modelling Study," *Science of the Total Environment* (2021): 148884.

61. Michael Cervin, The Truth About Bottled Water. Fox News. October 24, 2015. http://www.foxnews.com/health/2014/06/11/truth-about-bottled-water.html.

62. W. R. Osterkamp, *Annotated Definitions of Selected Geomorphic Terms*, USGS Open-File Report 2008–1217. (2008).

63. *Oral History of Floyd E. Dominy*, US Bureau of Reclamation Oral History Program, 168. https://www.usbr.gov/history/OralHistories/DOMINYMASTER3_2011.pdf.

64. David Brower, personal communication, 1990.

65. Anonymous, "California's David Brower, No. 1 Conservationist," *Life Magazine*, May 27, 1966, 37–38.

66. For a biography of David Brower see R. Wyss, *The Man Who Built the Sierra Club: A Life of David Brower* (New York: Columbia University Press, 2016); also see Brower's autobiography, David Brower, *For Earth's Sake: The Life and Times of David Brower* (Layton, UT: Peregrine-Smith Books, 1990).

67. The US Bureau of Reclamation maintains a trove of historical documents and project histories. These materials provide a great starting point for additional information about the Bureau. See https://www.usbr.gov/history/index.html.

68. For more on water transfers see F. Molle, "Cities versus Agriculture: Revisiting Intersectoral Water Transfers, Potential Gains, and Conflicts, vol. 10," *International Water Management Institute*, 2006; F. A. Schoolmaster, "Water Marketing and Water Rights Transfers in the Lower Rio Grande Valley, Texas," *Professional Geographer* 43, no. 3 (1991): 292–304; B. C. Saliba, "Do Water Markets 'Work'? Market Transfers and Trade-Offs in the Southwestern States," *Water Resources Research* 23, no. 7 (1987): 1113–22; and Charles W. Howe, Jeffrey K. Lazo, and Kenneth R. Weber, "The Economic Impacts of Agriculture-to-Urban Water Transfers on the Area of Origin: A Case Study of the Arkansas River Valley in Colorado," *American Journal of Agricultural Economics* 72, no. 5 (1990): 1200–1204.

69. For more about water development in California including the California State Water Project see M. Arax, *The Dreamt Land: Chasing Water and Dust across California* (Visalia, CA: Vintage, 2019).

70. Although acequias are physically equivalent to ditches, they possess unique governance institutions by virtue of their Hispanic origins and their singular importance to the communities where they operate. Consequently, acequias merit a separate entity in this book.

71. For decades geologists have studied and debated the origins of western canyons. For example, see I. Lucchitta, S. S. Bues, and M. Morales, "History of the Grand Canyon and of the Colorado River in Arizona: Geologic Evolution of Arizona," *Arizona Geological Society Digest* 17 (1989): 701–715; W. D. Lowry and E. M. Baldwin, "Late Cenozoic Geology of the Lower Columbia River Valley, Oregon, and Washington," *Geological Society of America Bulletin* 63, no. 1 (1952): 1–24; C. A. Ruleman, A. M. Hudson, R. A. Thompson, D. P. Miggins, J. B. Paces, and B. M Goehring, "Middle Pleistocene Formation of the Rio Grande Gorge, San Luis Valley, South-Central Colorado and North-Central New Mexico, USA: Process, Timing, and Downstream Implications," *Quaternary Science Reviews* 223 (2019), 105846; and W. R. Hansen, "Development of the Green River Drainage System Across the Uinta Mountains," *Geologic Guidebook of the Uinta Mountains: Utah's Maverick Range, Sixteenth Annual Field Conference* (1969): 93–100.

72. D. D. Jackson, ed., *Letters of the Lewis and Clark Expedition, with Related Documents, 1783–1854* (vol. 1) (Champaign: University of Illinois Press, 1978).

73. Editorial, "Grand Canyon 'Cash Registers,'" *Life Magazine*, May 7, 1965, 4.

74. The Navajo Generating Station ceased operations in 2019 as prices for natural gas and renewables such as solar power made further operation of the plant uneconomical. Workers began razing the plant in 2020. https://www.greentechmedia.com/articles/read/navajo-generating-station-coal-plant-closes-renewables, accessed October 19, 2020.

75. W. M. Hanemann, *The Central Arizona Project*, Department of Agricultural and Resource Economics, UCB. CUDARE Working Papers, Paper No. 937 (Berkeley: University of California, 2002).

76. For more about the Central Arizona Project see R. Johnson, *The Central Arizona Project: 1918–1968* (Tucson: University of Arizona Press, 1977).

77. US Bureau of Reclamation, *The Central Valley Project*, Bureau of Reclamation History Program, 1994. Retrieved 2018-07-09. For more detailed project histories see: https://www.usbr.gov/history/projhist.html#C.

78. See Congressional Research Service, July 2, 2020, *Central Valley Project: Issues and Legislation*. https://fas.org/sgp/crs/misc/R45342.pdf; see also Water Education Foundation, *Central Valley Project*. https://www.watereducation.org/aquapedia/central-valley-project.

79. Alex Simon, *Robert Towne: The Hollywood Interview* (2009). http://thehollywoodinterview.blogspot.com/2009/10/robert-towne-hollywood-interview.html. Accessed October 8, 2020.

80. For more about cienega hydrology and ecology see T. A. Minckley and A. Brunelle, "Paleohydrology and Growth of a Desert Ciénega," *Journal of Arid Environments* 69, no. 3 (2007): 420–431; see also L. E. Stevens and V. J. Meretsky, eds. *Aridland Springs in North America: Ecology and Conservation* (Tucson: University of Arizona Press, 2008).

81. One commonly sees references to "global warming" and "climate change" used interchangeably in the non-scientific literature. However, that use is technically incorrect, as global warming is just one subset of possible changes to the climate. For example, the climate might get warmer or colder in response to changes in atmospheric chemistry. Still, during the twentieth and twenty-first centuries, in most locations worldwide, the climate has become progressively hotter. Hence, we commonly see people use the term global warming when referring to the current state of climate change.

82. V. Masson-Delmotte, P. Zhai, A. Pirani, S. L. Connors, C. Péan, S. Berger, N. Caud, Y. Chen, L. Goldfarb, M. I. Gomis, M. Huang, K. Leitzell, E. Lonnoy, J. B. R. Matthews, T. K. Maycock, T. Waterfield, O. Yelekçi, R. Yu, and B. Zhou, eds., IPCC, 2021: *Climate Change 2021: The Physical Science Basis. Contribution of Working Group I to the Sixth Assessment Report of the Intergovernmental Panel on Climate Change* (Cambridge: Cambridge University Press, 2021).

83. A wide array of reports and information on global warming is available on the United Nations Intergovernmental Panel on Climate Change website. See https://www.ipcc.ch.

84. P. C. D. Milly and K. A. Dunne, "Colorado River Flow Dwindles as Warming-Driven Loss of Reflective Snow Energizes Evaporation," *Science* 367, no. 6483 (Mar. 13, 2020): 1252–55.

85. For a good overview on global warming and its effect on the American West, see William deBuys, *A Great Aridness: Climate Change and the Future of the American Southwest* (Oxford: Oxford University Press, 2012).

86. More specifically, the San Luis Valley Project was originally authorized in the Reclamation Project Act of 1939. The major feature of that project was the Platoro Dam. Then, in 1972, Congress authorized the Closed Basin Project as an additional element of the San Luis Valley Project. Construction on the Closed Basin Project began in 1980 and finished in the early 1990s.

87. For the project history see William Joe Simonds, *San Luis Valley Project*, US Bureau of Reclamation, 1994. https://www.usbr.gov/projects/pdf.php?id=187.

88. S. A. Changnon Jr. and J. L. Ivens, "History Repeated: The Forgotten Hail Cannons of Europe," *Bulletin of the American Meteorological Society* 62, no. 3 (1981): 368–75.

89. V. J. Schaefer, "The Production of Ice Crystals in a Cloud of Supercooled Water Droplets," *Science* 104, 457–59. (1946). https://doi.org/10.1126/science.104.2707.457; Bernard Vonnegut, "The Nucleation of Ice Formation by Silver Iodide," *Journal of Applied Physics* 18, no. 7 (1947): 593–95; Sarah A. Tessendorf, Jeffrey R. French, Katja Friedrich, Bart Geerts, Robert M. Rauber, Roy M. Rasmussen, Lulin Xue, et al., "A Transformational Approach to Winter Orographic Weather Modification Research: The SNOWIE Project," *Bulletin of the American Meteorological Society* 100, no. 1 (2019): 71–92.

90. For some history and basic principles of cloud seeding see W. R. Cotton and R. A. Pielke, *The Rise and Fall of the Science of Weather Modification by Cloud Seeding* (Cambridge: Cambridge University Press, 2007).

91. Robert M. Rauber, Bart Geerts, Lulin Xue, Jeffrey French, Katja Friedrich, Roy M. Rasmussen, Sarah A. Tessendorf, Derek R. Blestrud, Melvin L. Kunkel, and Shaun Parkinson, "Wintertime Orographic Cloud Seeding—A Review," *Journal of Applied Meteorology and Climatology* 58, no. 10 (2019): 2117–2140.

92. Katja Friedrich, Kyoko Ikeda, Sarah A. Tessendorf, Jeffrey R. French, Robert M. Rauber, Bart Geerts, Lulin Xue, et al., "Quantifying Snowfall from Orographic Cloud Seeding," *Proceedings of the National Academy of Sciences* 117, no. 10 (2020): 5190–5195.

93. Central Arizona Project, *Colorado River Basin Weather Modification Program*, March 1, 2018. http://www.cap-az.com/documents/meetings/2018-03-01/1683-4d-Web-Final-Action-Brief-CRC-Weather-Modification.pdf.

94. For more on the development of the Colorado System of Prior Appropriation and *Coffin v. Left Hand* in particular, see Robert R. Crifasi, *A Land Made from Water: Appropriation and the Evolution of Colorado's Landscape, Ditches, and Water Institutions* (Boulder: University Press of Colorado, 2015); David Schorr, *The Colorado Doctrine: Water Rights, Corporations, and Distributive Justice on the American Frontier* (New Haven: Yale University Press, 2012); and Coffin v. Left Hand Ditch Company, 6 Colo. 443 (1882).

95. For an environmental history of the Colorado River, see April R. Summit, *Contested Waters: An Environmental History of the Colorado River* (Boulder: University Press of Colorado, 2012); also Philip L. Fradkin, *A River No More: The Colorado River and the West* (Oakland: University of California Press, 1996).

96. Aldo Leopold, *A Sand County Almanac* (Oxford: Oxford University Press, 1949), 132.

97. Joseph C. Ives, *Report upon the Colorado River of the West*, US Government Printing Office, 1861, 28.

98. For more on the Colorado River Delta see E. R. Ward, *Border Oasis: Water and the Political Ecology of the Colorado River Delta, 1940–1975* (Tucson: University of Arizona Press, 2003).

99. Richard White, *The Organic Machine: The Remaking of the Columbia River* (New York: Hill and Wang, 1995).

100. J. Harlen Bretz, "The Channeled Scabland of the Columbia Plateau," *Journal of Geology* 31, no. 8 (1923): 617–649; J. Harlen Bretz, "The Spokane Flood Beyond the Channeled Scablands," *Journal of Geology* 33, no. 2 (1925): 97–115, 236; J. E. O'Connor and V. R. Baker, "Magnitudes and Implications of Peak Discharges from Glacial Lake Missoula," *Geological Society of America Bulletin* 104 (1992): 267–279.

101. John Wesley Powell, *The Exploration of the Colorado River and Its Canyons* (Mineola, NY: Dover Publications, 1961), 212.

102. J. N. Corbridge, T. A. Rice, and G. Vranesh, *Vranesh's Colorado Water Law* (Boulder: University Press of Colorado, 1999), 313.

103. Ed Quillen, "What Size Shoe Does an Acre-Foot Wear?," *High Country News*, November 10, 1986, 16–25.

104. Garrett Hardin's "Tragedy of the Commons" is indeed an influential work. According to Google Scholar, researchers have cited the article more than 45,000 times. See Garrett Hardin, "The Tragedy of the Commons," *Science* 162 (1968): 1243–1248.

105. Elinor Ostrom, "Tragedy of the Commons," *New Palgrave Dictionary of Economics* 2 (2008); Elinor Ostrom, "A General Framework for Analyzing Sustainability of Social-Ecological Systems," *Science* 325, no. 5939 (2009): 419–22; Elinor Ostrom, Joanna Burger, Christopher B. Field, Richard B. Norgaard, and David Policansky, "Revisiting the Commons: Local Lessons, Global Challenges," *Science* 284, no. 5412 (1999): 278–282; and Elinor Ostrom, "Elinor Ostrom Speaks about Property Rights," *Journal of Private Enterprise* 35, no. Fall 2020 (2020): 7–12.

106. Why hasn't the US gone metric? *Slate Magazine*. http://www.slate.com/articles/news_and_politics/explainer/1999/10/why_hasnt_the_us_gone_metric.html. Last accessed October 28, 2020.

107. John McPhee, "The Control of Nature: Los Angeles Against the Mountains–I," *New Yorker*, September 19, 1988.

108. Solifluction is an odd event that geomorphologists define as a slow, viscous downslope movement of waterlogged soil and sediment that is underlain by an impervious layer. You can best observe this event in arctic regions.

109. Lahars are another odd but particularly devastating hillslope failure known from volcanic areas where ash and water mix to form an unstoppable slurry that obliterates anything in its path.

110. Jane Kay, "Delta Smelt Icon of California Water Is Almost Extinct," *National Geographic*, April 3, 2015. https://news.nationalgeographic.com/2015/04/150403-smelt-california-bay-delta-extinction-endangered-species-drought-fish/; California Department of Fish and Wildlife, Delta Smelt. https://www.wildlife.ca.gov/Conservation/Fishes/Delta-Smelt. Last accessed October 28, 2020.

111. Charles W. Stockton, "Long-term Streamflow Reconstruction in the Upper Colorado River Basin Using Tree Rings," in Calvin G. Clyde, Donna H. Falkenborg, and J. Paul Riley, *Colorado River Basin Modeling Studies: Proceedings of a Seminar Held at Utah State University Logan, Utah, July 16–18, 1975*. Reports. Paper 538 (1976), 416. For more on how the Stockton chart was prepared see D. M. Meko, C. A. Woodhouse, C. H. Baisan, T. Knight, J. J. Lukas, M. K. Hughes, and M. W. Salzer, "Medieval Drought in the Upper Colorado River Basin," *Geophysical Research Letters* 34m (2007) L10705, doi: 10.1029/2007GL029988; for the original Stockton chart, see C. W. Stockton and G. C. Jacoby, "Long-term Surface-water Supply and Streamflow Trends in the Upper Colorado River Basin," *Lake Powell Research Project Bulletin* 18, National Science Foundation, 1976.

112. For more on the use of tree ring data on drought and water availability see J. J. Shinker, B. N. Shuman, T. A. Minckley, and A. K. Henderson, "Climatic Shifts in the Availability of Contested Waters: A Long-term Perspective from the Headwaters of the North Platte River," *Annals of the Association of American Geographers* 100, no. 4 (2010): 866–879.

113. Osborn and Caywood Ditch Co. v. Green, 673 P.2d 380, 383 (Colo. App. 1983).

114. Osborn and Caywood Ditch Co. v. Green, 673 P.2d 380, 383 (Colo. App. 1983).

115. For more on prescriptive rights see David H. Getches, *Water Law in a Nutshell* (Eagan, MN: Thomson West, 2009).

116. David Brower, *For Earth's Sake, the Life and Times of David Brower* (Layton, UT: Peregrine-Smith Books, 1990), 341.

117. "F. E. Dominy, Who Harnessed Water in the American West, is Dead at 100," *New York Times*, April 29, 2010, B13.

118. For an oral history of Floyd Dominy, see Brit Allan Storey, *Floyd Dominy Interview*, US Bureau of Reclamation Oral History Program. https://www.usbr.gov/history/OralHistories/DOMINYMASTER3_2011.pdf.

119. US Geological Survey, Water Science School, *Water Dowsing*. https://www.usgs.gov/special-topic/water-science-school/science/water-dowsing?qt-science_center_objects=0#qt-science_center_objects.

120. US Geological Survey. *Water Dowsing*, 10. https://pubs.usgs.gov/gip/water_dowsing/pdf/water_dowsing.pdf.

121. Scientists have demonstrated in the laboratory that water flowing through the microscopic channels of a ceramic filter can generate a small electric current. I am unaware of any research translating these results to groundwater systems. However, this research might provide a plausible mechanism for currents within groundwater. See, Jun Yang, Fuzhi Lu, Larry W. Kostiuk, and Daniel Y. Kwok, "Electrokinetic microchannel battery by means of electrokinetic and microfluidic phenomena," *Journal of Micromechanics and Microengineering* 13, no. 6 (2003): 963.

122. Sandra Postel, Paul Polak, Fernando Gonzales, and Jack Keller, "Drip Irrigation for Small Farmers: A New Initiative to Alleviate Hunger and Poverty," *Water International* 26, no. 1 (2001): 3–13.

123. US Drought Monitor. https://droughtmonitor.unl.edu/About/WhatistheUSDM.aspx.

124. Data that the US Drought Monitor uses to compile the drought severity maps include Palmer Drought Severity Index, the Standardized Precipitation Index, and other climatological inputs; the Keech-Byram Drought Index for fire; satellite-based assessments of vegetation health; and various indicators of soil moisture and hydrologic data, particularly in the West, such as the Surface Water Supply Index and snowpack.

125. For a popular account of the Dust Bowl, see Timothy Egan, *The Worst Hard Time: The Untold Story of Those Who Survived the Great American Dust Bowl* (Boston: Houghton Mifflin Harcourt, 2006); see also Donald Worster, *Dust Bowl: The Southern Plains in the 1930s* (Oxford: Oxford University Press, 2004); and Siegfried D. Schubert, Max J. Suarez, Philip J. Pegion, Randal D. Koster, and Julio T. Bacmeister, "On the Cause of the 1930s Dust Bowl," *Science* 303, no. 5665 (2004): 1855–1859.

126. John Steinbeck, *The Grapes of Wrath* (New York: Viking, 1939).

127. For more on the Echo Park struggles see chap. 12 in Roderick Nash, *Wilderness and the American Mind* (New Haven, CT: Yale University Press, 1982); Wallace Stegner, ed., *This Is Dinosaur: Echo Park Country and Its Magic Rivers* (New York: Alfred Knopf, 1955); and Debra Elaine Jenson, *Dinosaur Dammed: An Analysis of the Fight to Defeat Echo Park Dam* (PhD diss., University of Utah, 2014).

128. "Great Elephant Butte Dam Nearing Completion," *Popular Mechanics Magazine* 23, no. 1 (January 1915): 12–13.

129. National Park Service, *Elephant Butte Dam and Spillway New Mexico*. https://www.nps.gov/articles/new-mexico-elephant-butte-dam-and-spillway.htm. Last accessed October 28, 2020; for a good overview of Rio Grande controversies, litigation, and legislation, see Susan Kelly, Iris Augusten, Joshua Mann, and Lara Katz, "History of the Rio Grande Reservoirs in New Mexico: Legislation and Litigation 47," *Nat. Resources J.* 525 (2007). Available at https://digitalrepository.unm.edu/nrj/vol47/iss3/5.

130. For more on the Elwha Dam removal, see Tara Lohan, "The Elwha's Living laboratory: Lessons from the World's Largest Dam-removal Project," *The Revelator*, October 1, 2018. https://therevelator.org/elwha-dam-removal/; see also US National Park Service, *Elwha River Restoration*. https://www.nps.gov/olym/learn/nature/elwha-ecosystem-restoration.htm.

131. For a history of the environmental movement, see Roderick Nash, *Wilderness and the American Mind* (New Haven, CT: Yale University Press, 1982).

132. Nixon, facing a certain override by Congress should he veto the ESA, decided to sign the legislation and move on.

133. US Fish and Wildlife Service, *Laws and Policies*. https://www.fws.gov/endangered/laws-policies/.

134. For more on evapotranspiration, see M. E. Jensen, R. D. Burman, and R. G. Allen, "Evapotranspiration and Irrigation Water Requirements," *ASCE*, 1990; G. H. Hargreaves, and Z. A. Samani, "Reference Crop Evapotranspiration from Temperature," *Applied Engineering in Agriculture* 1, no. 2 (1985): 96–99; and G. H. Hargreaves and Z. A. Samani, "Estimating Potential Evapotranspiration," *Journal of the Irrigation and Drainage Division* 108, no. 3 (1982): 225–230.

135. Mark Fiege, *Irrigated Eden: The Making of an Agricultural Landscape in the American West* (Seattle: University of Washington Press, 1999).

136. US Geological Survey, *Water Science Glossary of Terms*. http://water.usgs.gov/edu/dictionary.html.

137. D. F. Ritter, R. C. Kochel, and J. R. Miller, *Process Geomorphology* (Dubuque, IA: Wm. C. Brown, 1995).

138. National Geographic Society, "Floodplain." https://www.nationalgeographic.org/encyclopedia/flood-plain/.

139. For detailed explanations of these terms and their differences, see A. D. Ward, A. D. Jayakaran, D. E. Mecklenburg, G. E. Powell, and J. Witter, "Two-stage Channel Geometry: Active Floodplain Requirements," *Encyclopedia of Water Science* (2008): 1253–1260.

140. For more on flood and flood planning see Thomas Dunne and Luna B. Leopold, *Water in Environmental Planning* (New York: W. H. Freeman and Company, 1978); moreover, the work of geographer Gilbert White transformed flood planning in the United States. See G. F. White, "Human Adjustment to Floods," *Department of Geography Research Paper No. 29* (University of Chicago, 1945). For more on the Black Hills flood, see US Geological Survey, The 1972 Black Hills–Rapid City Flood. https://www.usgs.gov/centers/dakota-water/science/1972-black-hills-rapid-city-flood?qt-science_center_objects=0#; see also National Weather Service, The Black Hills Flood of 1972. https://www.weather.gov/unr/1972-06-09. Last accessed October 19, 2020.

141. See National Institute of Dental and Craniofacial Research, *The Story of Fluoridation*. http://www.nidcr.nih.gov/oralhealth/Topics/Fluoride/TheStoryofFluoridation.htm.

142. For more on groundwater depletion and fossil water see L. F. Konikow and E. Kendy, "Groundwater Depletion: A Global Problem," *Hydrogeology Journal* 13 no. 1 (2005): 317–320; and L. F. Konikow, "Long-term Groundwater Depletion in the United States," *Groundwater* 53, no. 1 (2015): 2–9.

143. A. Kondash and A. Vengosh, "Water Footprint of Hydraulic Fracturing," *Environmental Science and Technology Letters* 2, no. 10 (2015): 276–280.

144. For an introduction to stream gaging see US Geological Survey, *Streamgaging Basics*. https://www.usgs.gov/mission-areas/water-resources/science/streamgaging-basics?qt-science_center_objects=0#qt-science_center_objects.

145. In 1888, USGS Director John Wesley Powell hired Frederick H. Newell as the first full-time employee of its new Irrigation Survey. One of Newell's tasks included developing standard water measurement methods. The USGS still uses many of Newell's methods today. Eventually, Newell became the first chief hydrographer of the USGS. Historians at the USGS believe that Newell adopted the USGS spelling of "gage" instead of "gauge." Newell apparently reasoned that *gage* was the proper Saxon spelling before the Norman influence added a 'u'. Newell might have also utilized the spelling of *gage* used in the standard dictionary (the first dictionary produced by Funk and Wagnalls). See pp. 28 and 50 in Robert Follansbee, *A History of the Water Resources Branch, U.S. Geological Survey: Volume 1 from Predecessor Surveys to June 30, 1919*, US Geological Survey, 1919.

146. Robert Follansbee, *A History of the Water Resources Branch, U.S. Geological Survey: Volume 1 from Predecessor Surveys to June 30* (US Geological Survey, 1919), 31.

147. Robert Follansbee, *A History of the Water Resources Branch, U.S. Geological Survey: Volume 1 from Predecessor Surveys to June 30* (US Geological Survey, 1919), 66.

148. For a biography of Gilbert see Harold L. Burstyn, *Grove Karl Gilbert: Explorer and Scientist* (US Geological Survey Open File Report 84–037, 1984).

149. Henry Faul and Carol Faul, *It Began with a Stone, A History of Geology from the Stone Age to the Age of Plate Tectonics* (Hoboken, NJ: John Wiley and Sons, 1983), 142–143. https://nsidc.org/cryosphere/glaciers/questions/what.html. Last accessed October 6, 2020.

150. For a reprint of Agassiz's 1840 monograph see Louise Agassiz, *Etudes sur les glaciers* (Cambridge: Cambridge University Press, 2012).

151. Perhaps most notoriously, Agassiz commissioned a series of photographs in 1850 of North Carolina slaves for his study of "races." Agassiz used these so-called "Zealy Daguerreotypes" (after the photographer who made them) to support his contention that nonwhite races were inferior. This tragic history illustrates how the production of factual knowledge is sometimes distorted by the persons who created it. For more on the Zealy Daguerreotypes, see Molly Rogers, *Delia's Tears: Race, Science, and Photography in Nineteenth-century America* (New Haven, CT: Yale University Press, 2010); for a more general biography of Agassiz before the extent of his racism became apparent, see Edward Lurie, *Louis Agassiz: A Life in Science* (Baltimore: Johns Hopkins University Press, 1988).

152. See http://sites.coloradocollege.edu/rockies/2013/09/05/the-effects-of-a-warming-climate-on-glacier-national-park/. For more about the South Cascade Glacier, see *Fifty-Year Record of Glacier Change Reveals Shifting Climate in the Pacific Northwest and Alaska*, USA, USGS Fact Sheet 2009–3046. http:// pubs.usgs.gov/fs/2009/3046/index.html.

153. David Brower in Eliot Porter, *The Place No One Knew: Glen Canyon on the Colorado* (San Francisco: Sierra Club–Ballantine, 1966).

154. Bureau of Reclamation, Glen Canyon Unit. https://www.usbr.gov/uc/rm/crsp/gc/.

155. For the best overview that I have read of these issues see part 3, "A River," in John McPhee, *Encounters with the Archdruid: Narratives About a Conservationist and Three of His Natural Enemies* (New York: Farrar, Straus and Giroux, 1977).

156. See, for example, D. P. Beard, *Deadbeat Dams: Why We Should Abolish the Bureau of Reclamation and Tear Down Glen Canyon Dam* (Chicago: Johnson Books, 2015); for a highly personal account of Glen Canyon and what it meant to one person, see Katie Lee,

All My Rivers Are Gone: A Journey of Discovery Through Glen Canyon (Boulder: Big Earth Publishing, 1998).

157. N. S. Christensen, A. W. Wood, N. Voisin, D. P. Lettenmaier, and R. N. Palmer, "The Effects of Climate Change on the Hydrology and Water Resources of the Colorado River Basin," *Climatic Change* 62, nos. 1–3 (2004): 337–363.

158. For a review of factors affecting reservoir operations, see S. J. Burges and R. K. Linsley, "Some Factors Influencing Required Reservoir Storage," *J. Hydraulics Division, ASCE* 97, no. HY7 (1971): 977–991.

159. Brit Allan Storey, *Floyd E. Dominy Oral Interview.* US Bureau of Reclamation Oral History Program. https://www.usbr.gov/history/OralHistories/DOMINYMASTER3_2011.pdf.

160. For a project history of the Columbia Basin Project, see William Joseph Simonds, *The Columbia Basin Project*, US Bureau of Reclamation, History Program, 1998.

161. Edwin James, *Account of an Expedition from Pittsburgh to the Rocky Mountains Performed in the Years 1819, 1820. 3 volumes* (London: Longman, Hurst, Rees, Orme, and Brown, 1823), vol. 3, 126.

162. For more on reclaimed water and its reuse see Metcalf and Eddy, Inc., Takashi Asano, Franklin L. Burton, Harold Leverenz, Ryujiro Tsuchihashi, and George Tchobanoglous, *Water Reuse* (New York: McGraw-Hill Professional Publishing, 2007); and James Crook and Rao Y. Surampalli, "Water Reclamation and Reuse Criteria in the US," *Water Science and Technology* 33, no. 10–11 (1996): 451–462.

163. For more on groundwater see F. G. Driscoll, *Groundwater and Wells, 2nd ed.* (St. Paul, MN: Johnson Division, 1986); and James A. Miller, ed., *Ground Water Atlas of the United States*, US Geological Survey, Denver, 2000.

164. For a biography of Carl Hayden see J. L. August, *Vision in the Desert: Carl Hayden and Hydropolitics in the American Southwest* (Fort Worth: Texas Christian University Press, 1999).

165. John Muir, "The Hetch Hetchy Valley," *Sierra Club Bulletin* 6, no. 4 (January 1908).

166. For a good overview of the Hetch Hetchy controversy see chap. 9 in Roderick Nash, *Wilderness and the American Mind* (New Haven, CT: Yale University Press, 1982). For a more complete account see R. W. Righter, *The Battle Over Hetch Hetchy: America's Most Controversial Dam and the Birth of Modern Environmentalism* (Oxford: Oxford University Press, 2005).

167. See Brian Fagan, "Insider: Phoenix's Looming Water Crisis," *Archaeology* 64, no. 2 (2011). http://archive.archaeology.org/1103/insider/phoenix_water_crisis_hohokam.html; for more on the Hohokam and their canals, see William H. Doelle, "Hohokam Heritage: The Casa Grande Community," *Archaeology Southwest* 23, no. 4 (2009); also M. Kyle Woodson, "Re-drawing the Map of the Hohokam Canals in the Middle Gila River Valley," *Journal of Arizona Archaeology* 1, no. 1 (2010): 5–20.

168. I adapted this summary of Boulder Dam construction from William Joe Simonds, reedited by Brit Storey, *The Boulder Canyon Project: Hoover Dam*, US Bureau of Reclamation, 2009. https://www.usbr.gov/history/hoover.html; also M. Hiltzik, *Colossus: Hoover Dam and the Making of the American Century* (New York: Simon and Schuster, 2010).

169. Wallace E. Stegner, *Beyond the Hundredth Meridian: John Wesley Powell and the Second Opening of the West* (Boston: Houghton Mifflin Company, 1954), 214.

170. Wallace E. Stegner, *Beyond the Hundredth Meridian: John Wesley Powell and the Second Opening of the West* (Boston: Houghton Mifflin Company, 1954), 229.

171. J. W. Powell, *Report on the Lands of the Arid Region of the United States: With a More Detailed Account of the Lands of Utah. With Maps*, US Government Printing Office, 1879, 2. Although Powell was the most famous person to publicize the 100th meridian notion, he was not the first. As early as 1866, John Hanson Beale wrote in *The Undeveloped West* that he saw wasteland for eight hundred miles west of the 100th meridian, and that the wasteland stretched from British Columbia to Mexico. See Wallace E. Stegner, *John Wesley Powell and the Second Opening of the West* (Boston: Houghton Mifflin, 1954), 216.

172. Robert R. Crifasi, "The Political Ecology of Water Use and Development," *Water International* 27, no. 4 (2002): 492–503.

173. For more on hybrid freshwater ecosystems see Robert R. Crifasi, *A Land Made from Water: Appropriation and the Evolution of Colorado's Landscape, Ditches, and Water Institutions* (Boulder: University Press of Colorado, 2015).

174. In 1917, Grove Karl Gilbert published his classic monograph on hydraulic mining in California. Among the groundbreaking concepts that Gilbert pioneered in his report were using a watershed perspective for the investigation, describing the extreme example of anthropogenic sedimentation, developing a quantitative sediment budget, and developing the sediment-wave theory of debris transport. See G. K. Gilbert, *Hydraulic-mining Debris in the Sierra Nevada*, US Geological Survey Professional Paper No. 105 (US Government Printing Office, 1917); for a discussion of the importance of this work see L. A. James, J. D. Phillips, and S. A. Leccec, "A Centennial Tribute to G. K. Gilbert's Hydraulic Mining Débris in the Sierra Nevada," *Geomorphology* 294, 1 October (2017): 4–19.

175. Harold L. Burstyn, *Grove Karl Gilbert: Explorer and Scientist*, US Geological Survey Open File Report 84–037, 1984, 11.

176. These statistics include Alaska and Hawaii. Source National Hydropower Association. Hydro.org.

177. Winters v. United States—207 U.S. 564, 28 S. Ct. 207 (1908); quote in G. C. Coggins and C. H. Wilkinson, *Federal Public Land and Resource Law*, 2nd ed. (West Conshohocken, PA: Foundation Press, 1987), 369.

178. Thomas Dunne and Luna B. Leopold, *Water in Environmental Planning* (New York: W. H. Freeman and Company, 1978), 194.

179. David M. Evans, "The Denver Area Earthquakes and the Rocky Mountain Arsenal Disposal Well," *The Mountain Geologist* (1966); for more on injection induced earthquakes see William L. Ellsworth, "Injection-induced Earthquakes," *Science* 341, no. 6142 (2013).

180. For a survey of instream flow water rights in the western US, see Cynthia F. Coveli, Phillips Whitney, and Alyson Scott, "Update to a Survey of State Instream Flow Programs in the Western United States," *University of Denver Water Law Review* 20 (2016): 355.

181. Constitution Annotated, ArtI.S10.C3.3 Compacts Clause. https://constitution.congress.gov/browse/essay/artI-S10-C3-3/ALDE_00001106/. Last accessed October 10, 2020.

182. US Constitution, ArtI.S10.C3.3 Compacts Clause.

183. For more on interstate water compacts and equitable apportionment, see G. Hobbs, *The Public's Water Resource: Articles on Water Law, History, and Culture*, Continuing Legal Education in Colorado, 2007, 75.

184. Treaty between the United States of America and Mexico, February 3, 1944. *Utilization of Waters of the Colorado and Tijuana Rivers and of the Rio Grande*. https://www.ibwc.gov/Files/1944Treaty.pdf.

185. European earthworms provide a good example, even though they are not an aquatic species. After all, since the earliest days of European colonization in the Americas, farmers have introduced earthworms into their soil.

186. For discussions about the nativeness of species, see Robert R. Crifasi, "A Subspecies No More? A Mouse, Its Unstable Taxonomy, and Western Riparian Resource Conflict," *Cultural Geographies* 14, no. 4 (2007): 511–535; R. M. Anderson, *From Non-native "Weed" to Butterfly "Host": Knowledge, Place, and Belonging in Ecological Restoration* (MA Thesis, University of Washington, 2017); Charles R. Warren, "Perspectives on the 'Alien' versus 'Native' Species Debate: A Critique of Concepts, Language, and Practice," *Progress in Human Geography* 31, no. 4 (2007): 427–46; J. Keulartz and C. Van der Weele, "Framing and Reframing in Invasion Biology," *Configurations* 16, no 1 (2008): 93–115; J. Keulartz, "Boundary Work in Ecological Restoration," *Environmental Philosophy* 6, no 1 (2009): 35–55; H. A. Perkins, "Killing One Trout to Save Another: A Hegemonic Political Ecology with Its Biopolitical Basis in Yellowstone's Native Fish Conservation Plan," *Annals of the American Association of Geographers* 110, no. 5 (2020): 1–18.

187. Powell was careful to note exceptions to the general north-south rule of thumb, such as parts of the Northwest that receive very high amounts of annual rainfall.

188. US Fish and Wildlife Service, Region 1, Lahontan cutthroat trout (*Oncorhynchus clarki henshawi*) Recovery Plan. Prepared by Patrick D. Coffin and William F. Cowan (January, 1995). https://ecos.fws.gov/docs/recovery_plan/950130.pdf.

189. Grove Karl Gilbert, *Lake Bonneville*, US Geological Survey Monograph. V. 1. US Government Printing Office, 1890. Although Gilbert's description of the lake itself has stood the test of time, the understanding of the tectonic events that shaped the Great Basin has changed dramatically since Gilbert's day. Geologists now believe that the Great Basin developed as a response to plate tectonics and the crustal expansion of western North America during the Cenozoic Era.

190. For research on the Bonneville Flood, see H. E. Malde, "The Catastrophic Late Pleistocene Bonneville Flood in the Snake River Plain, Idaho," *US Geological Survey Professional Paper* no. 596 (1968): 1–52.

191. I draw heavily on a history of the London Bridge and Lake Havasu from information published by Lake Havasu City. See Lake Havasu City website: http://www.golakehavasu.com/about-us/lakehavasucity_history.aspx.

192. "Acclaimed Architect Runs into Skeptical Angelenos," *New York Times*, September 24, 2015, A21.

193. B. Gumprecht, *The Los Angeles River: Its Life, Death, and Possible Rebirth* (Baltimore: John Hopkins University Press, 2001).

194. For more information on the life and accomplishments of Elwood Mead, see James Robert Kluger, *Elwood Mead: Irrigation Engineer and Social Planner* (PhD diss., University of Arizona, 1970).

195. John Nichols, *The Milagro Beanfield War: A Novel* (New York: Macmillan, 2000).

196. Elizabeth A. Fenn, *Encounters at the Heart of the World: A History of the Mandan People* (New York: Hill and Wang, 2014), 3. I recommend Fenn's book to anyone wishing to gain a better understanding of the Missouri River and its Indigenous inhabitants within the broad sweep of history.

197. For an environmental history of the Missouri see Philip Vincent Scarpino, *Great River: An Environmental History of the Upper Mississippi, 1890–1950* (Columbia: University of Missouri Press, 1985).

198. This story has appeared several times; for instance see Marc Reisner, *Cadillac Desert: The American West and its Disappearing Water* (New York: Penguin, 1993), 58.

199. For more on Mulholland see Catherine Mulholland, *William Mulholland and the Rise of Los Angeles* (Oakland: University of California Press, 2000); see also Norris Hundley, Donald C. Jackson, and Jean Patterson, *Heavy Ground: William Mulholland and the St. Francis Dam Disaster*, vol. 8 (Oakland: University of California Press, 2016). Estimates for the deaths from the flood vary. The Museum of Ventura County, CA, reports 424 known dead. See https://venturamuseum.org/journal-flashback/the-st-francis-dams-death-toll/.

200. You can find the text of the National Environmental Policy Act (Pub. L. 91–190, 42 U.S.C. 4321–4347, January 1, 1970, as amended by Pub. L. 94–52, July 3, 1975, Pub. L. 94–83, August 9, 1975, and Pub. L. 97–258, § 4(b), Sept. 13, 1982) here: https://www.usbr.gov/gp/nepa/cip/nepa.pdf.

201. If you are interested in reading more about nonpoint pollution, see David Zaring, "Agriculture, Nonpoint Source Pollution, and Regulatory Control: The Clean Water Act's Bleak Present and Future," *Harvard Environmental Law Review* 20 (1996): 515.

202. Christo, *Over the River*. https://christojeanneclaude.net/projects/over-the-river.

203. For more on the Owens Valley controversy see William L. Kahrl, *Water and Power: The Conflict over Los Angeles Water Supply in the Owens Valley* (Oakland: University of California Press, 1983).

204. Meanders are the ubiquitous curves, bends, and loops seen in alluvial river channels.

205. W. R. Osterkamp, *Annotated Definitions of Selected Geomorphic Terms*, US Geological Survey Open-File Report 2008–1217, 2008. https://www.wou.edu/las/physci/taylor/g322/Osterkamp_2008_USGS_ofr20081217.pdf.

206. For more on Parker Dam and the controversies arising during its construction see Toni Rae Linenberger, *Parker-Davis Project*, US Bureau of Reclamation. History Division, 1997. https://www.usbr.gov/projects/pdf.php?id=153.

207. William Denison Lyman, *The Columbia River: Its History, Its Myths, Its Scenery, Its Commerce* (New York: G. Putnam's Sons, 1901), 4.

208. Anna V. Smith, "The Klamath River Now Has the Legal Rights of a Person," *High Country News*, Sept. 24, 2019. https://www.hcn.org/issues/51.18/tribal-affairs-the-klamath-river-now-has-the-legal-rights-of-a-person.

209. Peter R. Briere, "Playa, Playa Lake, Sabkha: Proposed Definitions for Old Terms," *Journal of Arid Environments* 45, no. 1 (2000): 1–7; W. R. Osterkamp and W. Wood

Warren, "Playa-lake Basins on the Southern High Plains of Texas and New Mexico: Part I. Hydrologic, Geomorphic, and Geologic Evidence for Their Development," *Geological Society of America Bulletin* 99, no. 2 (1987): 215–223.

210. W. R. Osterkamp, *Annotated Definitions of Selected Geomorphic Terms*, USGS Open-File Report 2008–1217, 2008.

211. For anyone wanting to learn more about John Wesley Powell, the three indispensable books include Wallace E. Stegner, *Beyond the Hundredth Meridian: John Wesley Powell and the Second Opening of the West* (New York: Houghton Mifflin Company, 1954); Donald Worster, *A River Running West: The Life of John Wesley Powell* (Oxford University Press, 2002); and William deBuys, *Seeing Things Whole: The Essential John Wesley Powell* (Washington, DC: Island Press, 2001). Additionally, to commemorate the sesquicentennial of Powell's expedition down the Colorado, an edited volume was prepared that provides new scholarship and insights on Powell's achievements: J. Robinson, D. McCool, and T. Minckley, eds., *Vision and Place: John Wesley Powell and Reimagining the Colorado River Basin* (Oakland: University of California Press, 2020); furthermore, to get a sense of how much the landscape has changed since Powell's time, see H. G. Stephens, *In the Footsteps of John Wesley Powell: An Album of Comparative Photographs of the Green and Colorado Rivers, 1871–72 and 1968* (Chicago: Johnson Books, 1987).

212. A man named Horace Day designed the raft. Day added a fourth tube to later models. For a first-hand account of Fremont's journey, see John C. Fremont, *Narrative of the Exploring Expedition to the Rocky Mountains in the Near 1842; and to Oregon and North California, in the Years 1843–44* (Syracuse: Hall and Dickinson, 1847).

213. Charles D. Wilber, *Great Valleys and Prairies of Nebraska and the Northwest* (Omaha: Daily Republican Printer, 1881), 68; Charles Robert Kutzleb, "Can Forests Bring Rain to the Plains?" *Forest History* (1971): 14–21.

214. My principal source for this is Charles Robert Kutzleb, *Rain Follows the Plow: The History of an Idea* (PhD diss., University of Colorado, Boulder, 1968); Powell quote, *Lands of the Arid Regions*, 91; charlatan quote at Kutzleb, *Rain Follows the Plow*, 115.

215. For more on rainwater harvesting see Deep Narayan Pandey, Anil K. Gupta, and David M. Anderson, "Rainwater Harvesting as an Adaptation to Climate Change," *Current Science* 85, no. 1 (2003): 46–59.

216. American Whitewater Association, International Scale of River Difficulty. https://www.americanwhitewater.org/content/Wiki/safety:start?#vi._international_scale_of_river_difficulty.

217. The official name for the Newlands Act is the "Reclamation Act of 1902." Ch. 1093, 32 STAT. 388, June 17, 1902.

218. Robert M. Hughes, "Recreational Fisheries in the USA: Economics, Management Strategies, and Ecological Threats," *Fisheries Science* 81, no. 1 (2015): 1–9.

219. Hughes, "Recreational Fisheries in the USA," 1–9.

220. John Loomis, "Economic Contribution to the Colorado Economy and Benefits to Visitors from Water-Based Recreation," *Colorado Water*, July–August (2018), 9. http://cwi.colostate.edu/Media/img/newsletters/2018/CW_35_4.pdf.

221. C. Gregory Crampton and Gloria G. Griffen, "The San Buenaventura, Mythical River of the West," *The Pacific Historical Review* (1956): 163–171.

222. In Mexico, the Rio Grande is also known as the Rio Bravo. For more on the Rio Grande see Paul Horgan, *Great River: The Rio Grande in North American History. Vol. 1, Indians and Spain. Vol. 2, Mexico and the United States.* Two vols. in one (Middletown, CT: Wesleyan University Press, 2013).

223. American Whitewater has done an excellent job summarizing river access issues around the American West. Unless noted otherwise, I have drawn principally from their river navigability and public rights summaries published on their web site to prepare this section. American Whitewater remains an excellent source for up-to-date information on river access issues.

224. Hitchings, 55 Cal. App. 3d at 570.

225. Cal. Const. Article X, section 4.

226. Richard Gast, "People v. Emmert: A Step Backward for Recreational Water Use in Colorado," *University of Colorado Law Review*. 52 (1980): 247; Travis H. Burns, "Floating on Uncharted Headwaters: A Look at the Laws Governing Recreational Access on Waters of the Intermountain West," *Wyo. L. Rev.* 5 (2005): 561.

227. Idaho Revised Statutes 36–1601.

228. City of Coeur D'Alene v. Michael L. and Jeannette G. Mackin, et al. and State Board of Land Commissioners, et al., 147 P.3d 75 (Idaho 2006).

229. Adobe Whitewater Club v. State Game Commission. New Mexico Supreme Court, No S-1-SC-3895, (March 2, 2022).

230. Montana Coalition for Stream Access v. Curran, 682 P.2d 163, 168 (Mont. 1984).

231. Mont. Code Ann. § 23-2-302 (1997).

232. Robert N. Lane, "The Remarkable Odyssey of Stream Access in Montana," *Pub. Land and Resources L. Rev.* 36 (2015): 69.

233. State v. Bunkowski, 503 P.2d 1435 (Nev. 1972).

234. American Whitewater reports that following streams have been declared navigable under the federal title test in Nevada: (1) Colorado River by Nev. Rev. Stat. § 537.010 (2007); (2) Virgin River, including sources confluent above St. Thomas, by Nev. Rev. Stat. § 537.020 (2007); and (3) Carson River at Carson City by Bunkowski, 503 P.2d at 1236. A fourth body of water, Winnemucca Lake, was declared a navigable body of water by statute in 1921, Nev. Rev. Stat. § 537.030 (2007), but this lake is now entirely dry. The Nevada Supreme Court noted in Bunkowski that the United States Court of Appeals for the Ninth Circuit has held that Lake Tahoe is navigable. 503 P.2d at 1238 (citing Davis v. United States, 185 F.2d 938, 942–43 (9th Cir. 1950)).

235. 2005 WL 1079391 at 10.

236. John Mukum Mbaku, The Public Right to Float through Private Property in Utah: Conatser v. Johnson, *J. Land Resources and Envtl. L.* 29 (2009): 201.

237. Day v. Armstrong, 362 P.2d 137, 143 (Wyo. 1961).

238. 2005 WL 1079391 at 10.

239. Kemp v. Putnam, 288 P.2d 837, 839 (Wash. 1955).

240. Norman Maclean, *A River Runs Through It* (Chicago: University of Chicago Press, 1976).

241. Roger Ebert, "Review: *A River Runs Through It*," https://www.rogerebert.com/reviews/a-river-runs-through-it-1992 (1992).

242. Mark Stannard, D. Ogle, L. Holzworth, J. Scianna, and E. Sunleaf, "History, Biology, Ecology, Suppression and Revegetation of Russian-olive Sites (*Elaeagnus angustifolia L.*)," USDA-National Resources Conservation Service, Boise, ID, USA. Plant Materials 47, 2002.

243. Salinity in water and soils is of great concern to municipalities and farmers in arid regions, so much so that the federal government has invested heavily in controlling the salts in rivers. See for example US Department of Agriculture, Natural Resource Conservation Service, *Colorado River Basin Salinity Control Program*. https://www.nrcs.usda.gov/wps/portal/nrcs/detail/national/programs/alphabetical/?&cid=stelprdb1044198. Last accessed November 1, 2020.

244. S. K. Campbell and V. L. Butler, "Archaeological Evidence for Resilience of Pacific Northwest Salmon Populations and the Socioecological System Over the Last ~7,500 Years," *Ecology and Society* 15, no. 1 (2010): 1–20.

245. Campbell and Butler, "Archaeological Evidence," 1–20.

246. See Andrew Gobin, "Billy Frank Jr., the Foremost Champion for Treaty Indian Fishing, Dies at 83," *Tulalip News*, May 6, 2014. https://www.tulalipnews.com/wp/tag/boldt-decision/. For the text of the Boldt decision see United States v. Washington, 384 F. Supp. 312 (W.D. Wash. 1974), aff'd, 520 F.2d 676 (9th Cir. 1975).

247. Find a Grave, "Captain Harry Guleke," https://www.findagrave.com/memorial/25416363/harry-guleke.

248. K. O. Odigie, D. Hardisty, J. B. Geraci, and T. W. Lyons, "Historical Fluxes of Toxic Trace Elements in the Salton Sea Basin," *AGUFM* (2019): GH42A-06. For the Salton Sea dust mitigation program see Imperial Irrigation District, The Salton Sea Air Quality Mitigation Program. https://saltonseaprogram.com/aqm/. Last accessed October 23, 2020.

249. Ryan C. Rowland, Doyle W. Stephens, Bruce Waddell, and David L. Naftz, "Selenium Contamination and Remediation at Stewart Lake Waterfowl Management Area and Ashley Creek, Middle Green River Basin, Utah," *US Geological Survey Fact Sheet* 031–03; Kenneth Tanji, André Läuchli, and Jewell Meyer, "Selenium in the San Joaquin Valley," *Environment: Science and Policy for Sustainable Development* 28, no. 6 (1986): 6–39.

250. To understand Smythe's perspectives see William Ellsworth Smythe, *The Conquest of Arid America* (New York: Harper, 1900); also Lawrence B. Lee, "William Ellsworth Smythe and the Irrigation Movement: A Reconsideration," *Pacific Historical Review* 41, no. 3 (1972): 289–311; and chap. 4 in Patricia Nelson Limerick, *Desert Passages: Encounters with the American Deserts* (Albuquerque: University of New Mexico Press, 1985).

251. For more on the SNOTEL program, see the US Department of Agriculture, Natural Resource Conservation Service, *SNOTEL Data Collection Network Fact Sheet*. https://www.nrcs.usda.gov/wps/portal/wcc/home/aboutUs/publications/.

252. US Fish and Wildlife Service, Region 2, *Final Recovery Plan Southwestern Willow Flycatcher (Empidonax traillii extimus)*. Southwestern Willow Flycatcher Recovery Team Technical Subgroup (August 2002); the US Fish and Wildlife Service's Arizona Ecological Services Division maintains extensive information about the southwestern willow flycatcher on its web page. https://www.fws.gov/southwest/es/arizona/southwes.htm.

253. Philip H. Burgi, Bruce M. Moyes, and Thomas W. Gamble, "Operation of Glen Canyon Dam Spillways-Summer 1983," *Proceedings of the Conference Water for Resource*

Development, HY Di v. /ASCE Coeur d'Alene, Idaho (August 14–17, 1984). https://www.usbr.gov/tsc/techreferences/hydraulics_lab/pubs/PAP/PAP-0714.pdf.

254. For a popular account of the flooding of 1983 see K. Fedarko, *The Emerald Mile: The Epic Story of the Fastest Ride in History through the Heart of the Grand Canyon* (New York: Simon and Schuster, 2014); also "Grand Canyon High Water, 1983: Glen Canyon Damn [*sic*] Nearly Busts!" YouTube Video, https://www.youtube.com/watch?v=VPcrccxcNsI.

255. Robert Smithson, "The Spiral Jetty," in Gyorgy Kepes, ed., *Arts of the Environment*, 1972.

256. For more on Robert Smithson see Nancy Holt, *The Writings of Robert Smithson: Essays with Illustrations* (New York University Press, 1979).

257. For more on springs see N. Kresic and Z. Stevanovic, eds., *Groundwater Hydrology of Springs: Engineering, Theory, Management and Sustainability* (Burlington, MA: Butterworth-Heinemann, 2009).

258. Marco Cantonati, Roderick J. Fensham, Lawrence E. Stevens, Reinhard Gerecke, Douglas S. Glazier, Nico Goldscheider, Robert L. Knight, John S. Richardson, Abraham E. Springer, and Klement Tockner, "Urgent Plea for Global Protection of Springs," *Conservation Biology* 35, no. 1 (Sep. 2020): 378–382

259. Robert R. Crifasi, "Reflections in a Stock Pond: Are Anthropogenically Derived Freshwater Ecosystems Natural, Artificial, or Something Else?" *Environmental Management* 36, no. 5 (2005): 625–39.

260. For a technical summary with some dramatic images, see Garnett P. Williams and Markley Gordon Wolman, *Downstream Effects of Dams on Alluvial Rivers*, USGS Professional Paper 1286, 1984.

261. W. L. Graf, "Fluvial Adjustment to the Spread of Tamarisk in the Colorado Plateau Region," *GSA Bulletin* 89, (1978): 1491–1501; Daniel L. Keller, Brian G. Laub, Paul Birdsey, and David J. Dean, "Effects of Flooding and Tamarisk Removal on Habitat for Sensitive Fish Species in the San Rafael River, Utah: Implications for Fish Habitat Enhancement and Future Restoration Efforts," *Environmental Management* 54, no. 3 (2014): 465–78.

262. Roland de Gouvenain, "Origin, History and Current Range of Salt Cedar in the US," *Proceedings of the Saltcedar Management Workshop* (University of California, Davis, Miscellaneous Publication 1 [1996]): 1–3.

263. Geologists identify two basic types of river terraces, *fill terraces* and *strath terraces*. Fill terraces result when a valley fills with sediment. Strath terraces form during valley erosion. Here I am primarily discussing strath terraces.

264. For more on the geomorphology and formation of terraces see D. F. Ritter, R. C. Kochel, and J. R., Miller, *Process Geomorphology* (Dubuque, IA: William C. Brown, 1995); More recent research on terrace formation suggests that their evolution is more complex than previously believed; see Gregory S. Hancock and Robert S. Anderson, "Numerical Modeling of Fluvial Strath-terrace Formation in Response to Oscillating Climate," *Geological Society of America Bulletin* 114, no. 9 (2002): 1131–42.

265. For more on the Teton Dam failure see H. B. Seed and J. M. Duncan, "The Failure of Teton Dam," *Engineering Geology* 24, no. 1–4 (1987): 173–205; for some dramatic footage, see "Teton Dam Disaster," YouTube Video, https://www.youtube.com/watch?v=nQoMyBg5h_A&app=desktop.

266. For more on thermal springs see G. A. Waring and R. R. Blankenship, *Thermal Springs of the United States and Other Countries: A Summary*, US Geological Survey Professional Paper No. 492, Washington, 1965.

267. For more on tie drives, see M. K. Young, D. Haire, and M. A. Bozek, "The Effect and Extent of Railroad Tie Drives in Streams of Southeastern Wyoming," *Western Journal of Applied Forestry* 9, no. 4 (1994): 125–130; C. M. Ruffing, M. D. Daniels, and K. A. Dwire, "Disturbance Legacies of Historic Tie-drives Persistently Alter Geomorphology and Large Wood Characteristics in Headwater Streams, Southeast Wyoming," *Geomorphology* 231 (2015): 1–14; William H. Wroten Jr., *The Railroad Tie Industry in the Central Rocky Mountain Region, 1867–1900* (PhD diss., University of Colorado, Boulder, 1956).

268. For more on tinajas see B. T. Brown and R. R. Johnson, "The Distribution of Bedrock Depressions (Tinajas) as Sources of Surface Water in Organ Pipe Cactus National Monument, Arizona," *Journal of the Arizona-Nevada Academy of Science* (1983): 61–68.

269. For more on trans mountain water diversions, see Vujica Yevjevich, "Water Diversions and Interbasin Transfers," *Water International* 26, no. 3 (2001): 342–348; Raphael J. Moses, "Transmountain Diversions of Water in Colorado," *Denv. J. Int'l L. and Pol'y* 6 (1976): 329; Gregory M. Silkensen, "Windy Gap: Transmountain Water Diversion and the Environmental Movement," *Technical Report* 61 (Colorado Water Resources Research Institute, 1994).

270. For perhaps the best overview issues surrounding turf grass, see Paul Robbins, *Lawn People: How Grasses, Weeds, and Chemicals Make Us Who We Are* (Philadelphia: Temple University Press, 2012).

271. Between 1852 and 1861, Horace Greeley's newspaper, *New York Daily Tribune*, published more than 425 dispatches from Karl Marx and Frederick Engels. See articles by Marx in *New York Daily Tribune*, https://www.marxists.org/archive/marx/works/subject/newspapers/new-york-tribune.htm.

272. On Union Colony see Gregory J. Hobbs and Michael Welsh, *Confluence: The Story of Greeley Water* (Denver: Jordan Design, 2020); James F. Willard, *The Union Colony at Greeley, Colorado, 1869–1871* (Denver: W.F. Robinson Printing Co., 1918).

273. US Army Corps of Engineers, "A Brief History," https://www.usace.army.mil/About/History/Brief-History-of-the-Corps/. Last accessed July 9, 2018.

274. John A. Allan, "Policy Responses to the Closure of Water Resources," in *Water Policy: Allocation and Management in Practice*, ed. P. Howsam and R. Carter (London: Chapman and Hall, 1996).

275. John A. Allan, "Virtual Water: A Strategic Resource," *Ground Water* 36, no. 4 (1998): 545–547.

276. Tasnuva Mahjabin, Alfonso Mejia, Seth Blumsack, and Caitlin Grady, "Integrating Embedded Resources and Network Analysis to Understand Food-energy-water Nexus in the US," *Science of the Total Environment* 709 (2020): 136153.

277. Arjen Y. Hoekstra and Pin Q. Hung, "Virtual Water Trade," *Proceedings of the International Expert Meeting on Virtual Water Trade* 12 (2003): 1–244.

278. Paolo D'Odorico, Joel Carr, Carole Dalin, Jampel Dell'Angelo, Megan Konar, Francesco Laio, Luca Ridolfi, et al., "Global Virtual Water Trade and the Hydrological

Cycle: Patterns, Drivers, and Socio-environmental Impacts," *Environmental Research Letters* 14, no. 5 (2019): 053001.

279. Arjen Y. Hoekstra, Ashok K. Chapagain, Mesfin M. Mekonnen, and Maite M. Aldaya. *The Water Footprint Assessment Manual: Setting the Global Standard*, Routledge, 2011.

280. Christopher M. Chini, Lucas A. Djehdian, William N. Lubega, and Ashlynn S. Stillwell, "Virtual Water Transfers of the US Electric Grid," *Nature Energy* 3, no. 12 (2018): 1115–1123.

281. Findley quote in Tom I. Romero, "Color of Water: Observations of a Brown Buffalo on Water Law and Policy in Ten Stanzas," *U. Miami Race and Soc. Just. L. Rev.* 1 (2011): 107.

282. George Sibley, "Glen Canyon: Using a Dam to Heal a River," *High Country News*, July 22, 1996.

283. John Fleck, Jfleck at Inkstain, "You Can Almost See Its Feet," May 1, 2015, http://www.inkstain.net/fleck/page/115/.

284. For information on water conservation see J. P. Heaney, W. DeOreo, P. Mayer, P. Lander, J. Harpring, L. Stadjuhar, B. Courtney, and L. Buhlig, "Nature of Residential Water Use and Effectiveness of Conservation Programs," *Boulder Community Network* (2005); P. Mayer, P. Lander, and D. T. Glenn, "Outdoor Water Efficiency Offers Large Potential Savings, but Research on Effectiveness Remains Scarce," *Journal of the American Water Works Association* 2 (2015): 61.

285. Dolores Hayden, *A Field Guide to Sprawl* (New York: W. W. Norton and Company, 2004).

286. The Pueblo people are understandably reticent about sharing information about their religious and ritual practices. Consequently, it is difficult for outsiders to describe or draw conclusions about the religious significance of various rock art symbols. However, it is well known that the ritual significance of water in Puebloan life as expressed in rock art exists in many other forms besides water glyphs. As noted elsewhere Avanyu is a water deity found across the Southwest. In addition, one often sees petroglyphs of dragonflies and other water-related insects, stepped petroglyphs representing clouds, symbols for rain and lightning, waterfowl, and other water-loving birds. For a comprehensive exploration of cup and channel petroglyphs see Michael L. Terlep, "A Spatial and Stylistic Analysis of Cup and Channel Petroglyphs from the Arizona Strip" (MA thesis, Northern Arizona University, 2012).

287. Nevada District Court CV-1204049.

288. It is easy to take for granted that most Americans can drink water from their tap. The history of how we achieved this feat is an interesting one. See Martin V. Melosi, *The Sanitary City: Environmental Services in Urban America from Colonial Times to the Present* (University of Pittsburgh Press, 2008).

289. For more on potable water treatment, see Simon A. Parsons and Bruce Jefferson, *Introduction to Potable Water Treatment Processes* (Hoboken, NJ: Blackwell Publishing, 2006).

290. Sewer systems can provide unique insights into our communities. See Kevin L. Enfinger and Paul S. Mitchell, "Sewer Sociology—San Diego Metropolitan Area," in *Pipelines 2009: Infrastructure's Hidden Assets* (2009): 292–306; for an overview of

wastewater treatment technology, see Nicholas P. Cheremisinoff, *Handbook of Water and Wastewater Treatment Technologies* (Oxford: Butterworth-Heinemann, 2001).

291. John Wesley Powell, *Report on the Lands of the Arid Region of the United States: With a More Detailed Account of the Lands of Utah. With Maps.* US Government Printing Office, 1879.

292. See the discussion on virtual water for more on commodified water and its implications for conflict.

293. For more on the history of water wheels see Norman Smith, *Man and Water: A History of Hydro-technology* (New York: Scribner, 1975).

294. For more on wells see F. G. Driscoll, *Groundwater and Wells, 2nd ed.* (St. Paul, MN: Johnson Division, 1986).

295. Judy D. Fretwell, John S. Williams, and Phillip J. Redman, *National Water Summary on Wetland Resources*, US Geological Survey Water-Supply Paper 2425. US Government Printing Office, 1996.

296. Paul Stanton Kibel, "Grasp on Water: A Natural Resource that Eludes NAFTA's Notion of Investment," *Ecology LQ* 34 (2007): 655, 660.

297. For more on the whooping crane and specifically the efforts to conserve endangered species in the key Platte River corridor, see David M. Freeman, *Implementing the Endangered Species Act on the Platte Basin Water Commons* (Boulder: University Press of Colorado, 2010).

298. For more on the geologic history of Yellowstone Lake see William Richard Keefer, *The Geologic Story of Yellowstone National Park*. Vol. 1347. US Government Printing Office, 1971; Robert B. Smith and Lee J. Siegel, *Windows into the Earth: The Geologic Story of Yellowstone and Grand Teton National Parks* (Oxford University Press, 2000).

299. Leo R. Beard, "Estimating Long-term Storage Requirements and Firm Yield of Rivers," *Gen. Assem. of Berkeley of Intern. Union of Geodesy and Geophysics*, 1964.

300. US patent number 2,604,359A filed on June 27, 1949, and granted July 22, 1952.

301. For more on Zybach and the development of central pivot irrigation see J. P. Foley, "Centre Pivot and Lateral Move Machines," *WATERpak* (2008): 195–220.

302. Sabrina Hill, "The Legacy of Colorado's Largest Wildfire," Denver Water (June 16, 2017): https://www.denverwater.org/tap/legacy-colorados-largest-wildfire.

303. A. Park Williams, Benjamin I. Cook, and Jason E. Smerdon, "Rapid Intensification of the Emerging Southwestern North American Megadrought in 2020–2021," *Nature Climate Change* 12, no. 3 (2022): 232–34.

BIBLIOGRAPHY

Abbey, Edward. *Desert Solitaire: A Season in the Wilderness*. New York: Ballantine Books, 1968.

Abbey, Edward. *Down the River (with Henry Thoreau and Other Friends)*. New York: Plume, 1991.

Abbey, Edward. *The Journey Home: Some Words in the Defense of the American West*. New York: E. P. Dutton, 1977.

Abbey, Edward. *The Monkey Wrench Gang*. Salt Lake City: Dream Garden Press, 1985.

Abbey, Edward. *A Voice Crying in the Wilderness (Vox Clamantis in Deserto): Notes from a Secret Journal*. New York: St. Martin's Griffin, 1989.

Agassiz, Louis. *Etudes sur les Glaciers*. Cambridge: Cambridge University Press, 2012.

American Whitewater Association. International Scale of River Difficulty. https://www.americanwhitewater.org/content/Wiki/safety:start?#vi._international_scale_of_river_difficulty.

Anderson, R. M. "From Non-Native 'Weed' to Butterfly 'Host': Knowledge, Place, and Belonging in Ecological Restoration." MA Thesis, University of Washington, 2017.

Arax, M. *The Dreamt Land: Chasing Water and Dust Across California*. New York: Vintage, 2019.

August, J. L. *Vision in the Desert: Carl Hayden and Hydropolitics in the American Southwest*. Fort Worth: Texas Christian University Press, 1999.

Avalanche.org. https://avalanche.org.

Bailey, Jonathan. *Rock Art: A Vision of a Vanishing Cultural Landscape*. Boulder, CO: Johnson Books, 2016.

Barthes, Roland. *Camera Lucida: Reflections on Photography*. New York: Macmillan, 1981.

Beard, Daniel P. *Deadbeat Dams: Why We Should Abolish the Bureau of Reclamation and Tear Down Glen Canyon Dam*. Boulder, CO: Johnson Books, 2015.

Beard, Leo R. *Estimating Long-Term Storage Requirements and Firm Yield of Rivers*.

Gen. Assem. of Berkeley of Intern. Union of Geodesy and Geophysics, 1964.
Bourgeon, Lauriane, Ariane Burke, and Thomas Higham. "Earliest Human Presence in North America Dated to the Last Glacial Maximum: New Radiocarbon Dates from Bluefish Caves, Canada." *Plos one* 12, no. 1 (2017): e0169486.
Bretz, J. Harlen. "The Channeled Scabland of the Columbia Plateau." *Journal of Geology* 31, no. 8 (1923): 617–49.
Bretz, J. Harlen. "The Spokane Flood Beyond the Channeled Scablands." *Journal of Geology* 33, no. 2 (1925): 97–115, 236.
Briere, Peter R. "Playa, Playa Lake, Sabkha: Proposed Definitions for Old Terms." *Journal of Arid Environments* 45, no. 1 (2000): 1–7.
Brower, David. *For Earth's Sake: The Life and Times of David Brower.* Salt Lake City: Peregrine-Smith Books, 1990.
Brown, B. T., and R. R. Johnson. "The Distribution of Bedrock Depressions (Tinajas) as Sources of Surface Water in Organ Pipe Cactus National Monument, Arizona." *Journal of the Arizona-Nevada Academy of Science* (1983): 61–68.
Burges, S. J., and R. K. Linsley. "Some Factors Influencing Required Reservoir Storage." *J. Hydraulics Division*, ASCE 97, no. HY7 (1971): 977–91.
Burgi, Philip H., Bruce M. Moyes, and Thomas W. Gamble. "Operation of Glen Canyon Dam Spillways-Summer 1983." *Proceedings of the Conference Water for Resource Development, HY Di v. /ASCE Coeur d'Alene, Idaho* (August 14–17, 1984).
Burns, Travis H. "Floating on Uncharted Headwaters: A Look at the Laws Governing Recreational Access on Waters of the Intermountain West." *Wyo. L. Rev.* 5 (2005): 561.
Burstyn, Harold L. *Grove Karl Gilbert: Explorer and Scientist.* US Geological Survey Open File Report 84–037, 1984.
Cahalan, J. M. *Edward Abbey: A Life*. Tucson: University of Arizona Press, 2003.
California Department of Fish and Wildlife. Delta Smelt. https://www.wildlife.ca.gov/Conservation/Fishes/Delta-Smelt. Last accessed October 28, 2020.
Campbell, S. K., and V. L. Butler. "Archaeological Evidence for Resilience of Pacific Northwest Salmon Populations and the Socioecological System over the Last ~7,500 Years." *Ecology and Society* 15, no. 1 (2010): 1–20.
Cantonati, Marco, Roderick J. Fensham, Lawrence E. Stevens, Reinhard Gerecke, Douglas S. Glazier, Nico Goldscheider, Robert L. Knight, John S. Richardson, Abraham E. Springer, and Klement Tockner. "Urgent Plea for Global Protection of Springs." *Conservation Biology* (2020).
Central Arizona Project. Colorado River Basin Weather Modification Program. March 1, 2018. http://www.cap-az.com/documents/meetings/2018-03-01/1683-4d-Web-Final-Action-Brief-CRC-Weather-Modification.pdf.
Cervin, Michael. The Truth About Bottled Water. Fox News. October 24, 2015. http://www.foxnews.com/health/2014/06/11/truth-about-bottled-water.html.
Changnon, S. A., Jr., and J. L. Ivens. "History Repeated: The Forgotten Hail Cannons of Europe." *Bulletin of the American Meteorological Society* 62, no. 3 (1981): 368–75.
Cheremisinoff, Nicholas P. *Handbook of Water and Wastewater Treatment Technologies.* Oxford: Butterworth-Heinemann, 2001.
Chini, Christopher M., Lucas A. Djehdian, William N. Lubega, and Ashlynn S. Stillwell.

"Virtual Water Transfers of the US Electric Grid." *Nature Energy* 3, no. 12 (2018): 1115–1123.

Christensen, Christensen, A. W. Wood, N. Voisin, D. P. Lettenmaier, and R. N. Palmer. "The Effects of Climate Change on the Hydrology and Water Resources of the Colorado River Basin." *Climatic Change* 62, nos. 1–3 (2004): 337–63.

Christo. *Over the River.* https://christojeanneclaude.net/projects/over-the-river.

Coffin v. Left Hand Ditch Company, 6 Colo. 443 (1882).

Coggins, G. C., and C. H. Wilkinson. *Federal Public Land and Resource Law.* 2nd ed. Foundation Press, 1987.

Colorado Avalanche Information Center. https://www.avalanche.state.co.us.

Colorado Division of Water Resources. https://dwr.colorado.gov/services/water-administration/augmentation-plans. Last accessed October 28, 2020.

Colorado Water, July–August (2018). http://cwi.colostate.edu/Media/img/newsletters/2018/CW_35_4.pdf.

Congressional Research Service. July 2, 2020. Central Valley Project: Issues and Legislation. https://fas.org/sgp/crs/misc/R45342.pdf.

Constitution Annotated, ArtI.S10.C3.3 Compacts Clause. https://constitution.congress.gov/browse/essay/artI-S10-C3-3/ALDE_00001106/. Last accessed October 10, 2020.

Corbridge, J. N., T. A. Rice, and G. Vranesh. *Vranesh's Colorado Water Law.* Boulder: University Press of Colorado, 1999.

Cotton, W. R., and R. A. Pielke. *The Rise and Fall of the Science of Weather Modification by Cloud Seeding.* Cambridge: Cambridge University Press, 2007.

Coveli, Cynthia F., Whitney Phillips, and Alyson Scott. "Update to a Survey of State Instream Flow Programs in the Western United States." *University of Denver Water Law Review* 20 (2016): 355.

Crampton, C. Gregory, and Gloria G. Griffen. "The San Buenaventura, Mythical River of the West." *Pacific Historical Review* (1956): 163–171.

Crawford, Stanley G. *Mayordomo: Chronicle of an Acequia in Northern New Mexico.* Albuquerque: University of New Mexico Press, 1993.

Crifasi, Robert R. *A Land Made from Water: Appropriation and the Evolution of Colorado's Landscape, Ditches, and Water Institutions.* Boulder: University Press of Colorado, 2015.

Crifasi, Robert R. "The Political Ecology of Water Use and Development." *Water International* 27, no. 4 (2002): 492–503.

Crifasi, Robert R. "Reflections in a Stock Pond: Are Anthropogenically Derived Freshwater Ecosystems Natural, Artificial, or Something Else?" *Environmental Management* 36, no. 5 (2005): 625–639.

Crifasi, Robert R. "A Subspecies No More? A Mouse, Its Unstable Taxonomy, and Western Riparian Resource Conflict." *Cultural Geographies* 14, no. 4 (2007): 511–35.

Cronon, William. *Nature's Metropolis: Chicago and the Great West.* New York: W. W. Norton and Company, 2009.

Crook, James, and Rao Y. Surampalli. "Water Reclamation and Reuse Criteria in the US." *Water Science and Technology* 33, no. 10–11 (1996): 451–62.

Crossland, Charlotte Benson. "Acequia Rights in Law and Tradition." *Journal of the Southwest* (1990): 278–87.
Crutchfield, James A. *Revolt at Taos, the New Mexican and Indian Insurrection of 1847.* Yardley, PA: Westholme Publishing, 2015.
deBuys, William. *A Great Aridness: Climate Change and the Future of the American Southwest.* Oxford: Oxford University Press, 2012.
deBuys, William. *Seeing Things Whole: The Essential John Wesley Powell.* Washington, DC: Island Press, 2001.
deBuys, William. "Stewart Udall, John Wesley Powell, and the Emergence of a National American Commons." In *Vision and Place: John Wesley Powell and Reimagining the Colorado River Basin*, edited by Jason Robinson, Daniel McCool, and Thomas Minckley. Berkeley: University of California Press, 2020.
deBuys, William, and A. Harris. *River of Traps.* Albuquerque: University of New Mexico Press, 1990.
de Gouvenain, Roland. "Origin, History and Current Range of Salt Cedar in the US." In *Proceedings of the Saltcedar Management Workshop*, 1–3. University of California, Davis, Miscellaneous Publication 1 (1996): 1–52.
Deverell, William. "Fighting Words: The Significance of the American West in the History of the United States." *Western Historical Quarterly* 25, no. 2 (1994): 185–206.
D'Odorico, Paolo, Joel Carr, Carole Dalin, Jampel Dell'Angelo, Megan Konar, Francesco Laio, Luca Ridolfi, et al. "Global Virtual Water Trade and the Hydrological Cycle: Patterns, Drivers, and Socio-environmental Impacts." *Environmental Research Letters* 14, no. 5 (2019): 053001.
Doelle, William H. "Hohokam Heritage: The Casa Grande Community." *Archaeology Southwest* 23, no. 4 (2009).
Dolin, Eric J. *Fur, Fortune, and Empire: The Epic History of the Fur Trade in America.* London: W. W. Norton, 2010.
Driscoll, F. G. *Groundwater and Wells. 2nd ed.* St. Paul, MN: Johnson Division, 1986.
Dunne, Thomas, and Luna B. Leopold. *Water in Environmental Planning.* W. H. Freeman and Company, 1978.
Ebert, Roger. Review of *A River Runs Through It.* https://www.rogerebert.com/reviews/a-river-runs-through-it-1992 (1992).
Egan, Timothy. *The Worst Hard Time: The Untold Story of Those Who Survived the Great American Dust Bowl.* New York: Houghton Mifflin Harcourt, 2006.
Ellsworth, William L. "Injection-Induced Earthquakes." *Science* 341, no. 6142 (2013).
Emmons, Samuel Franklin, Whitman Cross, and George Homans Eldridge. *Geology of the Denver Basin in Colorado.* USGS Monograph Vol. 27. Washington, DC: US Government Printing Office, 1896.
Enfinger, Kevin L., and Paul S. Mitchell. "Sewer Sociology—San Diego Metropolitan Area." In *Pipelines 2009: Infrastructure's Hidden Assets* (2009): 292–306.
Evans, David M. "The Denver Area Earthquakes and the Rocky Mountain Arsenal Disposal Well." *The Mountain Geologist* (1966).
Fagan, Brian. "Insider: Phoenix's Looming Water Crisis." *Archaeology* 64, no. 2 (2011). http://archive.archaeology.org/1103/insider/phoenix_water_crisis_hohokam.html.

Faul, Henry, and Carol Faul. *It Began with a Stone, a History of Geology from the Stone Age to the Age of Plate Tectonics*. Hoboken, NJ: John Wiley and Sons, 1983.
Fedarko, K. *The Emerald Mile: The Epic Story of the Fastest Ride in History through the Heart of the Grand Canyon*. New York: Simon and Schuster, 2014.
Fenn, Elizabeth A. *Encounters at the Heart of the World: A History of the Mandan People*. New York: Hill and Wang, 2014.
Fiege, Mark. *Irrigated Eden: The Making of an Agricultural Landscape in the American West*. Seattle: University of Washington Press, 1999.
Foley, J. P. "Centre Pivot and Lateral Move Machines." *WATERpak* (2008): 195–220.
Follansbee, Robert. *A History of the Water Resources Branch*, US Geological Survey: Volume 1 from Predecessor Surveys to June 30, 1919 (1919).
Foster, Melissa A., Robert S. Anderson, Harrison J. Gray, and Shannon A. Mahan. "Dating of River Terraces along Lefthand Creek, Western High Plains, Colorado, Reveals Punctuated Incision." *Geomorphology* 295 (2017): 176–90.
Fradkin, P. L. *A River No More: The Colorado River and the West*. Berkeley: University of California Press, 1996.
Freeman, David M. *Implementing the Endangered Species Act on the Platte Basin Water Commons*. Boulder: University Press of Colorado, 2010.
Fremont, John C. *Narrative of the Exploring Expedition to the Rocky Mountains in the Near 1842; and to Oregon and North California, in the Years 1843–44*. Syracuse: Hall and Dickinson, 1847.
Fretwell, Judy D., John S. Williams, and Phillip J. Redman. "National Water Summary on Wetland Resources." US Geological Survey Water-Supply Paper 2425. Washington, DC: US Government Printing Office, 1996.
Friedrich, Katja, Kyoko Ikeda, Sarah A. Tessendorf, Jeffrey R. French, Robert M. Rauber, Bart Geerts, Lulin Xue, et al. "Quantifying Snowfall from Orographic Cloud Seeding." *Proceedings of the National Academy of Sciences* 117, no. 10 (2020): 5190–95.
Gast, Richard. "*People v. Emmert*: A Step Backward for Recreational Water Use in Colorado." *University of Colorado Law Review* 52 (1980): 247.
Gates, J. S. "Ground Water in the Great Basin Part of the Basin and Range Province, Western Utah." In *Cenozoic Geology of Western Utah: Sites for Precious Metal and Hydrocarbon Accumulations*, 75–90. Salt Lake City: Utah Geological Association, 1987.
Getches, D. H. *Water Law in a Nutshell* (No. 346.0432 G394 2009). Saint Paul, MN: Thomson West, 2009.
Gilbert, Grove Karl. "Hydraulic-mining Debris in the Sierra Nevada." US Geological Survey Professional Paper No. 105. Washington, DC: US Government Printing Office, 1917.
Gilbert, Grove Karl. *Lake Bonneville*. US Geological Survey Monograph, Vol. 1. Washington, DC: US Government Printing Office, 1890.
Gobin, Andrew. "Billy Frank Jr., the Foremost Champion for Treaty Indian Fishing, Dies at 83." *Tulalip News*. May 6, 2014. https://www.tulalipnews.com/wp/tag/boldt-decision/.
Graf, W. L. "Fluvial Adjustment to the Spread of Tamarisk in the Colorado Plateau Region." *GSA Bulletin* 89 (1978): 1491–1501.

"Grand Canyon High Water, 1983: Glen Canyon Damn [*sic*] Nearly Busts!" YouTube video. https://www.youtube.com/watch?v=VPcrccxcNsI.
Gumprecht, B. *The Los Angeles River: Its Life, Death, and Possible Rebirth*. Baltimore: John Hopkins University Press, 2001.
Guthman, Julie. "Agrarian Dreams: The Paradox of Organic Farming." In *California* (Vol. 11). Berkeley: University of California Press, 2014.
Hancock, Gregory S., and Robert S. Anderson. "Numerical Modeling of Fluvial Strath-Terrace Formation in Response to Oscillating Climate." *Geological Society of America Bulletin* 114, no. 9 (2002): 1131–42.
Hanemann, W. M. "The Central Arizona Project. Department of Agricultural and Resource Economics, UCB." CUDARE Working Papers, Paper No. 937. University of California, Berkeley, 2002.
Hansen, W. R. "Development of the Green River Drainage System Across the Uinta Mountains." In *Geologic Guidebook of the Uinta Mountains: Utah's Maverick Range*, Sixteenth Annual Field Conference, 93–100. Salt Lake City: Utah Geological Association, 1969.
Hardin, Garrett. "The Tragedy of the Commons." *Science* 162 (1968): 1243–48.
Hargreaves, G. H., and Z. A. Samani. "Estimating Potential Evapotranspiration." *Journal of the Irrigation and Drainage Division* 108, no. 3 (1982): 225–30.
Hargreaves, G. H., and Z. A. Samani. "Reference Crop Evapotranspiration from Temperature." *Applied Engineering in Agriculture* 1, no. 2 (1985): 96–99.
Hayden, Dolores. *A Field Guide to Sprawl*. London: W. W. Norton and Company, 2004.
Heaney, J. P., W. DeOreo, P. Mayer, P. Lander, J. Harpring, L. Stadjuhar, B. Courtney, and L. Buhlig. *Nature of Residential Water Use and Effectiveness of Conservation Programs*. Boulder Community Network, 2005.
Hill, Sabrina. "The Legacy of Colorado's Largest Wildfire," Denver Water (June 16, 2017). https://www.denverwater.org/tap/legacy-colorados-largest-wildfire.
Hiltzik, M. *Colossus: Hoover Dam and the Making of the American Century*. New York: Simon and Schuster, 2010.
Hobbs, G. *The Public's Water Resource: Articles on Water Law, History, and Culture*. Continuing Legal Education in Colorado Inc., 2007.
Hobbs, Gregory J., and Michael Welsh. *The Story of Greeley Water*. Denver: Jordan Design, 2020.
Hoekstra, Arjen Y., and Pin Q. Hung. "Virtual Water Trade." In *Proceedings of the International Expert Meeting on Virtual Water Trade* 12 (2003): 1–244.
Hoekstra, Arjen Y., Ashok K. Chapagain, Mesfin M. Mekonnen, and Maite M. Aldaya. *The Water Footprint Assessment Manual: Setting the Global Standard*. Milton Park, UK: Routledge, 2011.
Holt, Nancy. *The Writings of Robert Smithson: Essays with Illustrations*. New York: New York University Press, 1979.
Horgan, Paul. *Great River: The Rio Grande in North American History*. Vol. 1, *Indians and Spain*. Vol. 2, *Mexico and the United States*. 2 vols. in one. Middletown, CT: Wesleyan University Press, 2013.

Howe, Charles W., Jeffrey K. Lazo, and Kenneth R. Weber. "The Economic Impacts of Agriculture-to-Urban Water Transfers on the Area of Origin: A Case Study of the Arkansas River Valley in Colorado." *American Journal of Agricultural Economics* 72, no. 5 (1990): 1200–1204.

Hughes, Robert M. "Recreational Fisheries in the USA: Economics, Management Strategies, and Ecological Threats." *Fisheries Science* 81, no. 1 (2015): 1–9.

Hundley, Norris, Donald C. Jackson, and Jean Patterson. *Heavy Ground: William Mulholland and the St. Francis Dam Disaster.* Vol. 8. Berkeley: University of California Press, 2016.

Imperial Irrigation District. The Salton Sea Air Quality Mitigation Program. https://saltonseaprogram.com/aqm/. Last accessed October 23, 2020.

Ives, Joseph C. *Report Upon the Colorado River of the West.* Washington, DC: US Government Printing Office, 1861.

Jackson, D. D., ed. *Letters of the Lewis and Clark Expedition, with Related Documents, 1783–1854*, vol. 1. Champaign: University of Illinois Press, 1978.

James, Edwin. *Account of an Expedition from Pittsburgh to the Rocky Mountains Performed in the Years 1819, 1820.* 3 vols. Longman, Hurst, Rees, Orme, and Brown, 1823.

James, L. A., J. D. Phillips, and S. A. Leccec. "A Centennial Tribute to G. K. Gilbert's Hydraulic Mining Débris in the Sierra Nevada." *Geomorphology* 294, no. 1 (October 2017): 4–19.

Jenson, Debra Elaine. "Dinosaur Dammed: An Analysis of the Fight to Defeat Echo Park Dam." PhD diss., University of Utah, 2014.

Jensen, M. E., R. D. Burman, and R. G. Allen. *Evapotranspiration and Irrigation Water Requirements.* Reston, VA: ASCE, 1990.

Johnson, R. *The Central Arizona Project: 1918–1968.* Tucson: University of Arizona Press, 1977.

Kahrl, William L. *Water and Power: The Conflict over Los Angeles Water Supply in the Owens Valley.* Berkeley: University of California Press, 1983.

Kay, Jane. "Delta Smelt Icon of California Water Is Almost Extinct." *National Geographic.* April 3, 2015. https://news.nationalgeographic.com/2015/04/150403-smelt-california-bay-delta-extinction-endangered-species-drought-fish/.

Keefer, William Richard. *The Geologic Story of Yellowstone National Park.* Vol. 1347. Washington, DC: US Government Printing Office, 1971.

Keller, Daniel L., Brian G. Laub, Paul Birdsey, and David J. Dean. "Effects of Flooding and Tamarisk Removal on Habitat for Sensitive Fish Species in the San Rafael River, Utah: Implications for Fish Habitat Enhancement and Future Restoration Efforts." *Environmental Management* 54, no. 3 (2014): 465–78.

Kelly, Susan, Iris Augusten, Joshua Mann, and Lara Katz. "History of the Rio Grande Reservoirs in New Mexico: Legislation and Litigation." *Nat. Resources J.* 47, no. 525 (2007). https://digitalrepository.unm.edu/nrj/vol47/iss3/5.

Keulartz, J. "Boundary Work in Ecological Restoration." *Environmental Philosophy* 6, no. 1 (2009): 35–55.

Keulartz, J., and C. Van der Weele. "Framing and Reframing in Invasion Biology." *Configurations* 16, no. 1 (2008): 93–115.

Kibel, Paul Stanton. “Grasp on Water: A Natural Resource that Eludes NAFTA’s Notion of Investment.” *Ecology Law Quarterly* 34 (2007): 655.
Kluger, James Robert. “Elwood Mead: Irrigation Engineer and Social Planner.” PhD diss., University of Arizona, 1970.
Kondash, A., and A. Vengosh. “Water Footprint of Hydraulic Fracturing.” *Environmental Science and Technology Letters* 2, no. 10 (2015): 276–80.
Konikow, L. F. “Long-Term Groundwater Depletion in the United States.” *Groundwater* 53, no. 1 (2015): 2–9.
Konikow, L. F., and E. Kendy. “Groundwater Depletion: A Global Problem.” *Hydrogeology Journal* 13, no.1 (2005): 317–20.
Kresic, N., and Z. Stevanovic, eds. *Groundwater Hydrology of Springs: Engineering, Theory, Management and Sustainability*. Oxford: Butterworth-Heinemann, 2009.
Kutzleb, Charles Robert. “Can Forests Bring Rain to the Plains?” *Forest History* (1971): 14–21.
Kutzleb, Charles Robert. “Rain Follows the Plow: The History of an Idea.” PhD diss., University of Colorado, Boulder, 1968.
Lake Havasu City. City History. http://www.golakehavasu.com/about-us/lakehavasu city_history.aspx.
Lane, Robert N. “The Remarkable Odyssey of Stream Access in Montana.” *Pub. Land and Resources L. Rev.* 36 (2015): 69.
Lee, Katie. *All My Rivers Are Gone: A Journey of Discovery through Glen Canyon*. Boulder, CO: Big Earth Publishing, 1998.
Lee, Lawrence B. “William Ellsworth Smythe and the Irrigation Movement: A Reconsideration.” *Pacific Historical Review* 41, no. 3 (1972): 289–311.
Leopold, Aldo. *A Sand County Almanac*. Oxford: Oxford University Press, 1949.
Life Magazine. Close Up. “California’s David Brower, No. 1 Conservationist.” May 27, 1966, 37.
Life Magazine. Editorial. “Grand Canyon ‘Cash Registers.’” May 7, 1965, 4.
Limerick, Patricia Nelson. *Desert Passages: Encounters with the American Deserts*. Albuquerque: University of New Mexico Press, 1985.
Limerick, Patricia N. *The Legacy of Conquest: The Unbroken Past of the American West*. New York: W. W. Norton and Company, 1987.
Limerick, Patricia N., Jeffery Hickey, and Richard DiNucci. *What’s in a Name? Nichols Hall*. Boulder: University Press of Colorado, 1987.
Linenberger, Toni Rae. Parker-Davis Project. US Bureau of Reclamation. History Division, 1997. https://www.usbr.gov/projects/pdf.php?id=153.
Lohan, Tara. “The Elwha’s Living Laboratory: Lessons from the World’s Largest Dam-removal Project.” *Revelator*, October 1, 2018. https://therevelator.org/elwha-dam-removal/.
Loomis, John. “Economic Contribution to the Colorado Economy and Benefits to Visitors from Water-Based Recreation.” *Colorado Water* (July–August 2018). http://cwi.colostate.edu/Media/img/newsletters/2018/CW_35_4.pdf.
Lowry, W. D., and E. M. Baldwin. “Late Cenozoic Geology of the Lower Columbia River Valley, Oregon and Washington.” *Geological Society of America Bulletin* 63, no. 1

(1952): 1–24.
Lucchitta, I., S. S. Bues, and M. Morales. “History of the Grand Canyon and of the Colorado River in Arizona.” *Geologic Evolution of Arizona: Arizona Geological Society Digest* 17 (1989): 701–15.
Lurie, Edward. *Louis Agassiz: A Life in Science*. Baltimore: Johns Hopkins University Press, 1988.
Lyman, William Denison. *The Columbia River: Its History, Its Myths, Its Scenery, Its Commerce*. New York: G. P. Putnam’s Sons, 1901.
Maclean, Norman. *A River Runs Through It*. Chicago: University of Chicago Press, 1976.
Mahjabin, Tasnuva, Alfonso Mejia, Seth Blumsack, and Caitlin Grady. “Integrating Embedded Resources and Network Analysis to Understand Food-Energy-Water Nexus in the US.” *Science of the Total Environment* 709 (2020): 136153.
Malde, H. E. “The Catastrophic Late Pleistocene Bonneville Flood in the Snake River Plain, Idaho.” US Geological Survey Professional Paper No. 596 (1968): 1–52.
Masson-Delmotte, V., P. Zhai, A. Pirani, S. L. Connors, C. Péan, S. Berger, N. Caud, Y. Chen, L. Goldfarb, M. I. Gomis, M. Huang, K. Leitzell, E. Lonnoy, J. B. R. Matthews, T. K. Maycock, T. Waterfield, O. Yelekçi, R. Yu, and B. Zhou, eds. *IPCC, 2021: Climate Change 2021: The Physical Science Basis. Contribution of Working Group I to the Sixth Assessment Report of the Intergovernmental Panel on Climate Change*. Cambridge: Cambridge University Press, 2021.
Mayer, P., P. Lander, and D. T. Glenn. “Outdoor Water Efficiency Offers Large Potential Savings, but Research on Effectiveness Remains Scarce.” *Journal of the American Water Works Association* 2 (2015): 61.
Mbaku, John Mukum. “The Public Right to Float through Private Property in Utah: *Conatser v. Johnson*.” *J. Land Resources and Envtl. L.* 29 (2009): 201.
McPhee, John. “The Control of Nature: Los Angeles against the Mountains—1.” *New Yorker*. September 19, 1988.
McPhee, John. *Encounters with the Archdruid: Narratives about a Conservationist and Three of His Natural Enemies*. New York: Farrar, Straus and Giroux, 1977.
Mead, Elwood, and C. T. Johnson. *Use of Water in Irrigation: Report of Investigations Made in 1899*. Bulletin 86, US Department of Agriculture, Office of Experiment Stations. Washington, DC: US Printing Office, 1901.
Meko, D. M., C. A. Woodhouse, C. H. Baisan, T. Knight, J. J. Lukas, M. K. Hughes, and M. W. Salzer. “Medieval Drought in the Upper Colorado River Basin.” *Geophysical Research Letters* 34m L10705 (2007). doi: 10.1029/2007GL029988.
Melosi, Martin V. *The Sanitary City: Environmental Services in Urban America from Colonial Times to the Present*. Pittsburgh: University of Pittsburgh Press, 2008.
Menand, Louis. “Out of Bethlehem, The Radicalization of Joan Didion.” *New Yorker*. August 17, 2015.
Metcalf and Eddy, Inc., Takashi Asano, Franklin L. Burton, Harold Leverenz, Ryujiro Tsuchihashi, and George Tchobanoglous. *Water Reuse*. New York: McGraw-Hill Professional Publishing, 2007.
Metropolitan Museum of Modern Art. “What Is Modern Art?” https://www.moma.org/learn/moma_learning/themes/what-is-modern-art/. Last accessed October 28,

2020.
Miller, James A., ed. *Ground Water Atlas of the United States*. Washington, DC: US Geological Survey, 2000.
Milly, P. C. D., and K. A. Dunne. "Colorado River Flow Dwindles as Warming-Driven Loss of Reflective Snow Energizes Evaporation." *Science* 367, no. 6483 (March 13, 2020): 1252–55.
Minckley, T. A., and A. Brunelle. "Paleohydrology and Growth of a Desert Ciénega." *Journal of Arid Environments* 69, no. 3 (2007): 420–31.
Molle, F., and J. Berkoff. "Cities Versus Agriculture: Revisiting Intersectoral Water Transfers, Potential Gains, and Conflicts." *International Water Management Institute* 10 (2006).
Moses, Raphael J. "Transmountain Diversions of Water in Colorado." *Denv. J. Int'l L. and Pol'y* 6 (1976): 329.
Muir, John. "The Hetch Hetchy Valley." *Sierra Club Bulletin* 6, no. 4 (January 1908).
Mulholland, Catherine. *William Mulholland and the Rise of Los Angeles*. Berkeley: University of California Press, 2000.
Nash, Roderick. *Wilderness and the American Mind*. New Haven, CT: Yale University Press, 1982.
National Environmental Policy Act. Pub. L. 91–190, 42 U.S.C. 4321–47, January 1, 1970, as amended by Pub. L. 94–52, July 3, 1975, Pub. L. 94–83, August 9, 1975, and Pub. L. 97–258, § 4(b), Sept. 13, 1982.
National Geographic Society. "Floodplain." https://www.nationalgeographic.org/encyclopedia/flood-plain/.
National Hydropower Association. Hydro.org.
National Institute of Dental and Craniofacial Research. "The Story of Fluoridation." http://www.nidcr.nih.gov/oralhealth/Topics/Fluoride/TheStoryofFluoridation.htm.
National Park Service. "Elephant Butte Dam and Spillway New Mexico." https://www.nps.gov/articles/new-mexico-elephant-butte-dam-and-spillway.htm. Last accessed October 28, 2020.
National Weather Service. "The Black Hills Flood of 1972." https://www.weather.gov/unr/1972-06-09. Last accessed October 19, 2020.
New York Times. "Acclaimed Architect Runs into Skeptical Angelenos," September 24, 2015, A21.
New York Times. "F. E. Dominy, Who Harnessed Water in the American West, Is Dead at 100," April 29, 2010, B13.
Nichols, John. *The Milagro Beanfield War: A Novel*. New York: Macmillan, 2000.
Norris, Ned, Jr. "Stop Destroying Tohono O'odham Sacred Sites." *High Country News*, December 2020: 43.
O'Connor, J. E., and V. R. Baker. "Magnitudes and Implications of Peak Discharges from Glacial Lake Missoula." *Geological Society of America Bulletin* 104 (1992): 267–79.
Odigie, K. O., D. Hardisty, J. B. Geraci, and T. W. Lyons. "Historical Fluxes of Toxic Trace Elements in the Salton Sea Basin." *AGUFM* (2019): GH42A-06.
Osborn and Caywood Ditch Co. v. Green, 673 P.2d 380, 383 (Colo. App. 1983).

Osterkamp, W. R. "Annotated Definitions of Selected Geomorphic Terms." USGS Open-File Report 2008–1217, 2008. https://www.wou.edu/las/physci/taylor/g322/Osterkamp_2008_USGS_ofr20081217.pdf.

Osterkamp, W. R., and Warren W. Wood. "Playa-lake Basins on the Southern High Plains of Texas and New Mexico: Part I. Hydrologic, Geomorphic, and Geologic Evidence for Their Development." *Geological Society of America Bulletin* 99, no. 2 (1987): 215–23.

Ostrom, Elinor. "Elinor Ostrom Speaks about Property Rights." *Journal of Private Enterprise* 35 (Fall 2020): 7–12.

Ostrom, Elinor. "A General Framework for Analyzing Sustainability of Social-ecological Systems." *Science* 325, no. 5939 (2009): 419–22.

Ostrom, Elinor. "Tragedy of the Commons." *The New Palgrave Dictionary of Economics* 2 (2008).

Ostrom, Elinor, Joanna Burger, Christopher B. Field, Richard B. Norgaard, and David Policansky. "Revisiting the Commons: Local Lessons, Global Challenges." *Science* 284, no. 5412 (1999): 278–82.

Owen, David. "The World Is Running Out of Sand." *New Yorker*, May 29, 2017.

Pandey, Deep Narayan, Anil K. Gupta, and David M. Anderson. "Rainwater Harvesting as an Adaptation to Climate Change." *Current Science* 85, no. 1 (2003): 46–59.

Parsons, Simon A., and Bruce Jefferson. *Introduction to Potable Water Treatment Processes*. Hoboken, NJ: Blackwell Publishing, 2006.

Patten, D. T., L. Rouse, and J. C. Stromberg. "Isolated Spring Wetlands in the Great Basin and Mojave Deserts, USA: Potential Response of Vegetation to Groundwater Withdrawal." *Environmental Management* 41, no. 3 (2008): 398–413.

Perkins, H. A. "Killing One Trout to Save Another: A Hegemonic Political Ecology with Its Biopolitical Basis in Yellowstone's Native Fish Conservation Plan." *Annals of the American Association of Geographers* 110, no. 5 (2020): 1–18.

Popular Mechanics Magazine. "Great Elephant Butte Dam Nearing Completion." 23, no. 1 (January 1915): 12–13.

Porter, Eliot. *In Wildness Is the Preservation of the World*. Oakland, CA: Sierra Club, 1962.

Porter, Eliot. *The Place No One Knew: Glen Canyon on the Colorado*. Oakland, CA: Sierra Club, 1966.

Postel, Sandra, Paul Polak, Fernando Gonzales, and Jack Keller. "Drip Irrigation for Small Farmers: A New Initiative to Alleviate Hunger and Poverty." *Water International* 26, no. 1 (2001): 3–13.

Powell, John Wesley. *The Exploration of the Colorado River and Its Canyons*. Mineola, NY: Dover Publications, 1961.

Powell, John Wesley. *Report on the Lands of the Arid Region of the United States: With a More Detailed Account of the Lands of Utah. With Maps*. Washington, DC: US Government Printing Office, 1879.

Prudden, T. Mitchell. "The Prehistoric Ruins of the San Juan Watershed in Utah, Arizona, Colorado, and New Mexico." *American Anthropologist* 5, no. 2 (1903): 224–88.

Quillen, Ed. "What Size Shoe Does an Acre-Foot Wear?" *High Country News*, November 10, 1986, 16–25.

Rauber, Robert M., Bart Geerts, Lulin Xue, Jeffrey French, Katja Friedrich, Roy M. Rasmussen, Sarah A. Tessendorf, Derek R. Blestrud, Melvin L. Kunkel, and Shaun Parkinson. "Wintertime Orographic Cloud Seeding—A Review." *Journal of Applied Meteorology and Climatology* 58, no. 10 (2019): 2117–2140.

Reclamation Act of 1902. Ch. 1093, 32 STAT. 388, June 17, 1902.

Reisner, Marc. *Cadillac Desert: The American West and Its Disappearing Water.* New York: Penguin, 1993.

Righter, R. W. *The Battle Over Hetch Hetchy: America's Most Controversial Dam and the Birth of Modern Environmentalism.* Oxford: Oxford University Press, 2005.

Ritter, D. F., R. C. Kochel, and J. R. Miller. *Process Geomorphology.* Dubuque, IA: Wm. C. Brown, 1995.

Robbins, Paul. *Lawn People: How Grasses, Weeds, and Chemicals Make Us Who We Are.* Philadelphia: Temple University Press, 2012.

Robinson, J., D. McCool, and T. Minckley, eds. *Vision and Place: John Wesley Powell and Reimagining the Colorado River Basin.* Berkeley: University of California Press, 2020.

Rodríguez, Sylvia. *Acequia: Water Sharing, Sanctity, and Place.* Santa Fe, NM: School for Advanced Research Resident Scholar Book, 2006.

Rogers, Molly, and David W. Blight. *Delia's Tears: Race, Science, and Photography in Nineteenth-century America.* New Haven, CT: Yale University Press, 2010.

Romero, Tom I. "Color of Water: Observations of a Brown Buffalo on Water Law and Policy in Ten Stanzas." *U. Miami Race & Soc. Just. L. Rev.* 1 (2011): 107.

Rowland, Ryan C., Doyle W. Stephens, Bruce Waddell, and David L. Naftz. "Selenium Contamination and Remediation at Stewart Lake Waterfowl Management Area and Ashley Creek, Middle Green River Basin, Utah." US Geological Survey Fact Sheet 031–03.

Ruffing, C. M., M. D. Daniels, and K. A. Dwire. "Disturbance Legacies of Historic Tie-Drives Persistently Alter Geomorphology and Large Wood Characteristics in Headwater Streams, Southeast Wyoming." *Geomorphology* 231 (2015): 1–14.

Ruleman, C. A., A. M. Hudson, R. A. Thompson, D. P. Miggins, J. B. Paces, and B. M. Goehring. "Middle Pleistocene Formation of the Rio Grande Gorge, San Luis Valley, South-Central Colorado and North-Central New Mexico, USA: Process, Timing, and Downstream Implications." *Quaternary Science Reviews* 223 (2019): 105846.

Saliba, B. C. "Do Water Markets 'Work'? Market Transfers and Trade-Offs in the Southwestern States." *Water Resources Research* 23, no. 7 (1987): 1113–22.

Scarpino, Philip Vincent. *Great River: An Environmental History of the Upper Mississippi, 1890–1950.* Columbia: University of Missouri Press, 1985.

Schaefer, V. J. "The Production of Ice Crystals in a Cloud of Supercooled Water Droplets." *Science* 104 (1946): 457–59. https://doi.org/10.1126/science.104.2707.457.

Schoolmaster, F. A. "Water Marketing and Water Rights Transfers in the Lower Rio Grande Valley, Texas." *Professional Geographer* 43, no. 3 (1991): 292–304.

Schorr, David. *The Colorado Doctrine: Water Rights, Corporations, and Distributive Justice on the American Frontier.* New Haven, CT: Yale University Press, 2012.

Schubert, Siegfried D., Max J. Suarez, Philip J. Pegion, Randal D. Koster, and Julio T. Bacmeister. "On the Cause of the 1930s Dust Bowl." *Science* 303, no. 5665 (2004): 1855–59.

Schulte, S. C. *Wayne Aspinall and the Shaping of the American West*. Boulder: University Press of Colorado, 2002.
Seed, H. B., and J. M. Duncan. "The Failure of Teton Dam." *Engineering Geology* 24, nos. 1–4 (1987): 173–205.
Sheridan, Thomas E. "Arizona: Political Ecology of a Desert State." *Journal of Political Ecology* 2 (1995): 41–57.
Shinker, J. J., B. N. Shuman, T. A. Minckley, and A. K. Henderson. "Climatic Shifts in the Availability of Contested Waters: A Long-term Perspective from the Headwaters of the North Platte River." *Annals of the Association of American Geographers* 100, no. 4 (2010): 866–79.
Sibley, George. "Glen Canyon: Using a Dam to Heal a River." *High Country News*, July 22, 1996.
Silkensen, Gregory M. "Windy Gap: Transmountain Water Diversion and the Environmental Movement." Technical report no. 61. Colorado Water Resources Research Institute, 1994.
Simon, Alex. "Robert Towne: The Hollywood Interview." 2009. http://thehollywood interview.blogspot.com/2009/10/robert-towne-hollywood-interview.html, accessed October 8, 2020.
Simonds, William Joe. Reedited by Brit Storey, 2009. The Boulder Canyon Project: Hoover Dam, US Bureau of Reclamation, (2009). https://www.usbr.gov/history /hoover.html.
Simonds, William Joe. "The Columbia Basin Project." US Bureau of Reclamation, History Program, 1998. https://www.usbr.gov/projects/pdf.php?id=88.
Simonds, William Joe. "San Luis Valley Project." US Bureau of Reclamation, 1994. https://www.usbr.gov/projects/pdf.php?id=187.
Slate Magazine. "Why Hasn't the US Gone Metric?" http://www.slate.com/articles /news_and_politics/explainer/1999/10/why_hasnt_the_us_gone_metric.html. Last accessed October 28,2020.
Smith, Anna V. "The Klamath River Now Has the Legal Rights of a Person." *High Country News*, September 24, 2019. https://www.hcn.org/issues/51.18/tribal-affairs -the-klamath-river-now-has-the-legal-rights-of-a-person.
Smith, Norman. *Man and Water: A History of Hydro-Technology*. New York: Scribner, 1975.
Smith, Robert B., and Lee J. Siegel. *Windows into the Earth: The Geologic Story of Yellowstone and Grand Teton National Parks*. Oxford: Oxford University Press, 2000.
Smithson, Robert. "The Spiral Jetty." In *Arts of the Environment*. Edited by Gyorgy Kepes. New York: George Brazilleer, Inc., 1972.
Smythe, William Ellsworth. *The Conquest of Arid America*. New York: Macmillan, 1911.
Spicer, Edward H. *Cycles of Conquest: The Impact of Spain, Mexico, and the United States on the Indians of the Southwest, 1533–1960*. Tucson: University of Arizona Press, 1962.
Stannard, Mark, D. Ogle, L. Holzworth, J. Scianna, and E. Sunleaf. "History, Biology, Ecology, Suppression and Revegetation of Russian-Olive Sites (*Elaeagnus angustifolia L.*)." USDA-National Resources Conservation Service, Boise, ID, USA. *Plant Materials* 47 (2002).

Stegner, W. E. *Beyond the Hundredth Meridian: John Wesley Powell and the Second Opening of the West*. Boston: Houghton Mifflin Company, 1954.

Stegner, Wallace, ed. *This is Dinosaur: Echo Park and Its Magic Rivers*. New York: Alfred A. Knopf, 1955.

Steinbeck, John. *The Grapes of Wrath*. New York: Viking, 1939.

Stephens, H. G. *In the Footsteps of John Wesley Powell: An Album of Comparative Photographs of the Green and Colorado Rivers, 1871–72 and 1968*. Boulder, CO: Johnson Books, 1987.

Stevens, L. E., and V. J. Meretsky, eds. *Aridland Springs in North America: Ecology and Conservation*. Tucson: University of Arizona Press, 2008.

Stockton, Charles W. "Long-Term Streamflow Reconstruction in the Upper Colorado River Basin Using Tree Rings." In *Colorado River Basin Modeling Studies: Proceedings of a Seminar Held at Utah State University Logan, Utah, July 16–18, 1975*. Edited by Calvin G. Clyde, Donna H. Falkenborg, and J. Paul Riley. Reports. Paper 538. (1976).

Stockton, C. W., and G. C. Jacoby. "Long-Term Surface-Water Supply and Streamflow Trends in the Upper Colorado River Basin." Lake Powell Research Project Bulletin No. 18. National Science Foundation, 1976.

Storey, Brit Allan. "Dominy, Floyd E. Interview." US Bureau of Reclamation Oral History Program. https://www.usbr.gov/history/OralHistories/DOMINYMASTER3_2011.pdf.

Sturgeon, S. C. *The Politics of Western Water: The Congressional Career of Wayne Aspinall*. Tucson: University of Arizona Press, 2002.

Summit, April R. *Contested Waters: An Environmental History of the Colorado River*. Boulder: University Press of Colorado, 2012.

Swyngedouw, Erik. "Modernity and Hybridity: Nature, Regeneracionismo, and the Production of the Spanish Waterscape, 1890–1930." *Annals of the Association of American Geographers* 89, no. 3 (1999): 443–65. http://dx.doi.org/10.1111/0004-5608.00157.

Tanji, Kenneth, André Läuchli, and Jewell Meyer. "Selenium in the San Joaquin Valley." *Environment: Science and Policy for Sustainable Development* 28, no. 6 (1986): 6–39.

Terlep, Michael L. "A Spatial and Stylistic Analysis of Cup and Channel Petroglyphs from the Arizona Strip." MA thesis, Northern Arizona University, 2012.

Tessendorf, Sarah A., Jeffrey R. French, Katja Friedrich, Bart Geerts, Robert M. Rauber, Roy M. Rasmussen, Lulin Xue, et al. "A Transformational Approach to Winter Orographic Weather Modification Research: The SNOWIE Project." *Bulletin of the American Meteorological Society* 100, no. 1 (2019): 71–92.

"Teton Dam Disaster." YouTube video, https://www.youtube.com/watch?v=nQoMyBg5h_A&app=desktop.

Treaty between the United States of America and Mexico. Treaty of Guadalupe Hidalgo (1848) https://www.ourdocuments.gov/doc.php?flash=false&doc=26&page=transcript.

Treaty between the United States of America and Mexico, February 3, 1944. Utilization of Waters of the Colorado and Tijuana Rivers and of the Rio Grande. https://www.ibwc.gov/Files/1944Treaty.pdf.

United Nations Intergovernmental Panel on Climate Change Website. https://www.ipcc.ch.
US Army Corps of Engineers. "A Brief History." https://www.usace.army.mil/About/History/Brief-History-of-the-Corps/. Last accessed July 9, 2018.
US Bureau of Reclamation. "The Central Valley Project." Bureau of Reclamation History Program, 1994. Retrieved July 9, 2018.
US Bureau of Reclamation. "Project Histories Documents." https://www.usbr.gov/history/index.html.
US Department of Agriculture, Natural Resource Conservation Service. "Colorado River Basin Salinity Control Program." https://www.nrcs.usda.gov/wps/portal/nrcs/detail/national/programs/alphabetical/?&cid=stelprdb1044198. Last accessed November 1, 2020.
US Department of Agriculture, Natural Resource Conservation Service. "SNOTEL Data Collection Network Fact Sheet." http://www.wcc.nrcs.usda.gov/factpub/sntlfct1.html.
US Drought Monitor. https://droughtmonitor.unl.edu/About/WhatistheUSDM.aspx.
US Fish and Wildlife Service. Arizona Ecological Services Division. Southwestern Willow Flycatcher. https://www.fws.gov/southwest/es/arizona/southwes.htm.
US Fish and Wildlife Service. Laws and Policies. https://www.fws.gov/endangered/laws-policies/.
US Fish and Wildlife Service, Region 1, Lahontan cutthroat trout (Oncorhynchus clarki henshawi) Recovery Plan. Prepared by Patrick D. Coffin William F. Cowan (January, 1995). https://ecos.fws.gov/docs/recovery_plan/950130.pdf.
US Fish and Wildlife Service, Region 2, Final Recovery Plan Southwestern Willow Flycatcher (Empidonax traillii extimus). Southwestern Willow Flycatcher Recovery Team Technical Subgroup (August 2002). https://www.fws.gov/southwest/es/Documents/R2ES/Southwestern_Willow_Flycatcher_FINAL_Recovery_Plan_Aug_2002.pdf.
US Geological Survey. Fifty-Year Record of Glacier Change Reveals Shifting Climate in the Pacific Northwest and Alaska, USA. USGS Fact Sheet. 2009–3046. http:// pubs.usgs.gov /fs/2009/3046/index.html.
US Geological Survey. The 1972 Black Hills–Rapid City Flood. https://www.usgs.gov/centers/dakota-water/science/1972-black-hills-rapid-city-flood?qt-science_center_objects=0#.
US Geological Survey. Stream Gaging Basics. https://www.usgs.gov/mission-areas/water-resources/science/streamgaging-basics?qt-science_center_objects=0#qt-science_center_objects.
US Geological Survey. Water Dowsing. https://pubs.usgs.gov/gip/water_dowsing/pdf/water_dowsing.pdf.
US Geological Survey. Water Science Glossary of Terms. http://water.usgs.gov/edu/dictionary.html.
US Geological Survey. Water Science School. Water Dowsing. https://www.usgs.gov/special-topic/water-science-school/science/water-dowsing?qt-science_center_objects=0#qt-science_center_objects.

US National Park Service. Elwha River Restoration. https://www.nps.gov/olym/learn/nature/elwha-ecosystem-restoration.htm.
United States v. Washington, 384 F. Supp. 312 (W.D. Wash. 1974), aff'd, 520 F.2d 676 (9th Cir. 1975).
Villanueva, Cristina M., Marianna Garfí, Carles Milà, Sergio Olmos, Ivet Ferrer, and Cathryn Tonne. "Health and Environmental Impacts of Drinking Water Choices in Barcelona, Spain: A Modelling Study." *Science of the Total Environment* (2021): 148884.
Vonnegut, Bernard. "The Nucleation of Ice Formation by Silver Iodide." *Journal of Applied Physics* 18, no. 7 (1947): 593–95.
Walsh, Bryan. "Dying for a Drink." *Time*, December 4, 2008. http://content.time.com/time/magazine/article/0,9171,1864440,00.html. Accessed October 28, 2020.
Ward, A. D., A. D. Jayakaran, D. E. Mecklenburg, G. E. Powell, and J. Witter. "Two-Stage Channel Geometry: Active Floodplain Requirements." *Encyclopedia of Water Science* (2008): 1253–60.
Ward, E. R. *Border Oasis: Water and the Political Ecology of the Colorado River Delta, 1940–1975*. Tucson: University of Arizona Press, 2003.
Waring, G. A., and R. R. Blankenship. "Thermal Springs of the United States and Other Countries: A Summary." US Geological Survey Professional Paper No. 492. Washington, DC, 1965.
Warren, Charles R. "Perspectives on the 'Alien' Versus 'Native' Species Debate: A Critique of Concepts, Language and Practice." *Progress in Human Geography* 31, no. 4 (2007): 427–46.
Water Education Foundation. "Central Valley Project." https://www.watereducation.org/aquapedia/central-valley-project.
Weymouth, F. E. "Metropolitan Water District of Southern California History and First Annual Report." Metropolitan Water District of Southern California, 1938.
White, Gilbert F. "Human Adjustment to Floods." Department of Geography Research Paper No. 29. Chicago: University of Chicago, 1945.
White, Richard. *A New History of the American West*. Norman: University of Oklahoma Press, 1991
White, Richard. *The Organic Machine: The Remaking of the Columbia River*. New York: Hill and Wang, 1995.
Wilber, Charles D. *Great Valleys and Prairies of Nebraska and the Northwest*. Omaha: Daily Republican Printer, 1881.
Willard, James F. *The Union Colony at Greeley, Colorado, 1869–1871*. Denver: W. F. Willard Printing Co., 1918.
Williams, A. Park, Benjamin I. Cook, and Jason E. Smerdon. "Rapid Intensification of the Emerging Southwestern North American Megadrought in 2020–2021." *Nature Climate Change* 12, no. 3 (2022): 232–34.
Williams, Garnett P., and Markley Gordon Wolman. "Downstream Effects of Dams on Alluvial Rivers." USGS Professional Paper 1286. 1984.
Winters v. United States—207 U.S. 564, 28 S. Ct. 207 (1908).
Wohl, Ellen. *Virtual Rivers, Lessons from the Mountain Rivers of the Colorado Front Range*. New Haven, CT: Yale University Press, 2001.

Woodson, M. Kyle. "Re-drawing the Map of the Hohokam Canals in the Middle Gila River Valley." *Journal of Arizona Archaeology* 1, no. 1 (2010): 5–20.

Worster, Donald. *Dust Bowl: The Southern Plains in the 1930s*. Oxford: Oxford University Press, 2004.

Worster, Donald. *A River Running West: The Life of John Wesley Powell*. Oxford: Oxford University Press, 2002.

Worster, Donald. *Rivers of Empire: Water, Aridity, and the Growth of the American West*. Oxford: Oxford University Press, 1992.

Wroten, William H., Jr. "The Railroad Tie Industry in the Central Rocky Mountain Region 1867–1900." PhD diss., University of Colorado, Boulder, 1956.

Wyss, R. *The Man Who Built the Sierra Club: A Life of David Brower*. New York: Columbia University Press, 2016.

Yang, Jun, Fuzhi Lu, Larry W. Kostiuk, and Daniel Y. Kwok. "Electrokinetic microchannel battery by means of electrokinetic and microfluidic phenomena." *Journal of Micromechanics and Microengineering* 13, no. 6 (2003): 963.

Yevjevich, Vujica. "Water Diversions and Interbasin Transfers." *Water International* 26, no. 3 (2001): 342–48.

Young, M. Jane. *Signs from the Ancestors: Zuni Cultural Symbolism and Perceptions of Rock Art*. Albuquerque: University of New Mexico Press, 1988.

Young, M. K., D. Haire, and M. A. Bozek. "The Effect and Extent of Railroad Tie Drives in Streams of Southeastern Wyoming." *Western Journal of Applied Forestry* 9, no. 4 (1994): 125–30.

Zaring, David. "Agriculture, Nonpoint Source Pollution, and Regulatory Control: The Clean Water Act's Bleak Present and Future." *Harvard Environmental Law Review* 20 (1996): 515.

Zybach Self-Propelled Sprinkling Apparatus US patent number 2,604,359A. Filed on June 27, 1949, and granted July 22, 1952.

INDEX

Page numbers followed by *f* indicate illustrations. Page numbers followed by *n* indicate endnotes.